New Approaches in Foundation Engineering

Edited by Salih Yilmaz

Published in London, United Kingdom

New Approaches in Foundation Engineering
http://dx.doi.org/10.5772/intechopen.97974
Edited by Salih Yilmaz

Contributors
Jianhai Zhang, Tianzhi Yao, Li Qian, Zuguo Mo, Yunpeng Gao, Fujun Xue, Chenggang Liao, Zhong Zhou, Natalia Gotman, Suchit Kumar Patel, Jon Sinnreich, Steffen Leppla, Rolf Katzenbach, Maísa Comar Pinhotti Aguiar

First published in London, United Kingdom, 2022 by IntechOpen
IntechOpen is the global imprint of INTECHOPEN LIMITED, registered in England and Wales, registration number: 11086078, 5 Princes Gate Court, London, SW7 2QJ, United Kingdom

British Library Cataloguing-in-Publication Data
A catalogue record for this book is available from the British Library

Additional hard and PDF copies can be obtained from orders@intechopen.com

New Approaches in Foundation Engineering
Edited by Salih Yilmaz
p. cm.
Print ISBN 978-1-80355-429-7
Online ISBN 978-1-80355-430-3
eBook (PDF) ISBN 978-1-80355-431-0

Meet the editor

Dr. Salih Yilmaz received his Ph.D. in Civil Engineering from Pamukkale University, Turkey, in 2007. After serving as a research assistant and lecturer there, Dr. Yilmaz transferred to Katip Celebi University, Turkey, where he founded the Department of Civil Engineering and Earthquake Engineering Research Center in 2012 and 2014, respectively. He was a faculty member at both Katip Celebi University and Pamukkale University and a research associate at the University of Texas at Austin, USA. Dr. Yılmaz is involved in several national and international research projects and has consulted for many companies. He is currently a practicing structural and geotechnical engineer. His interests include foundation-based structural damages, seismic performance of existing buildings and geotechnical structures, soil improvement and foundation strengthening, and artificial intelligence applications in geotechnical engineering.

Contents

Preface

Foundation engineering is one of the most important parts of structural design and one of the most important subdisciplines in civil engineering. It is also related to many fields such as geology, geophysics, and mining engineering. In recent years, in addition to improvements in computer-aided design, new approaches in structural design have brought new perspectives into foundation engineering.

This book consists of two sections. Chapters in the first section discuss the determination and improvement of soil properties. Chapters in the second section discuss new approaches to foundation design. The book provides a new perspective on the analysis and assessment of soils as well as methods for their improvement. It also serves as an important guide for designers regarding sustainability and the use of artificial intelligence techniques in foundation design.

As one of the thousands of Turkish academics who were dismissed from their job with an unjust accusation given in one night about 6 years ago, I gladly accepted the book editorial offer to me by Intech. Because, in those days, it was almost impossible for me to engage in academic life in Turkey and I was just trying to survive by taking part in different positions in construction companies. I hope that you will benefit from the works presented by the authors of the book which will be a beginning step towards my new academic life in the days when I just left my homeland for freedom.

Salih Yilmaz
Asylum Seeker Academics,
Athens, Greece

Section 1

Soil Properties and Ground Improvement Methods

Chapter 1

Tropical Soils: Considerations on Occurrence and Characteristics and Studies in Brazil

Maisa Comar Pinhotti Aguiar

Abstract

Soil engineering is challenging as soils are found in different conditions of climate, topography and location and have properties that vary in depth and laterally. The tropical soils represent types of soils with similar characteristics and can be found in locations where tropical climate occurs. Inevitably, engineering projects must consider the particularities of these soils due to their genesis and the study of their properties, which requires investigations ranging from simpler methods to more complex field and laboratory tests, is essential to predict their behavior. In the case of lateritic soils, their differentiated characteristics impose additional challenges on designers since their classification does not fit perfectly into the classifications formulated for temperate climate soils. One limitation found in developing countries is that there are few or no geotechnical mappings and the knowledge of these soils are restricted at the central-south region of Brazil, where the major urban centres and the largest engineering companies and laboratories are located. Although studied by authors from all over the world, it is still possible to deepen the understanding of the properties of tropical soils and their application in the most diverse requests of civil engineering and others. This chapter was developed from bibliographic research.

Keywords: tropical soil, lateritic soils, Brazil, behavior

1. Introduction

Soils are formed by the joint action of climatic factors, weather, relief, source material, and organisms. Source materials consist of rocks or other soils, under which the other factors work. Thus, the properties of a residual soil and the behavior it presents in the face of various requests will largely be determined by the rock or material from which it originated.

In tropical and intertropical regions, a diversity of climates and relief is observed, resulting in a very large variety of soils known as tropical.

Brazil, for its large size (more than 8 million square kilometers), presents geological, climatic, and relief diversity that has conditioned the formation of soils with various behaviors.

In general, the significant climatic factors for soil formation are mean precipitation and temperature, which condition the rates of chemical reactions, the rate of change of rocks as well as the mobility of elements along the profile. Formed from the leaching of bases and the concentration of oxides and iron and aluminum sesquioxides in this pedological evolution, we have the lateritic soils, which have properties differentiated from the soils formed in temperate climate.

Thus, considering the wide distribution of tropical climate in the world and its occurrence in much of the Brazilian territory, aspects on the genesis, importance, and properties of tropical soils in the country will be addressed in this chapter.

2. Tropical soils and Brazil

The definition of tropical soil varies from region to region [1], but in general they are defined as those that occur in places that have tropical and humid climates.

The tropics are regions of the Earth located approximately in the middle of the globe between the latitude lines of the Tropic of Cancer and the Tropic of Capricorn and include the Ecuador line and parts of North America, South America, Africa, Asia, and Australia; tropical regions are home to about one-third of the world's population and account for 36% of the land mass.

Intemperism in the tropics can reach tens of meters below the surface, and the products of this process are complex and are not only of interest to geotechnical engineers; they are of great interest to other researchers [2]. It is possible to say that tropical soils are rather intemperated soils rich in iron oxides and aluminum; however, not all tropical soils can be included in this category, since they can originate from materials such as volcanic gray or form in regions of desert climate and thus exhibit different characteristics of the indicated [3].

In this sense, Brazil has 92% of Brazil's territory located in the Tropical or Intertropical Climate Zone, the remaining 8% are south of the Tropic of Capricorn and are inserted in the Temperate Climate Zone of the Southern Hemisphere. (**Figure 1**) in which the climate Aw is observed (the tropical savanna climate features distinct wet and dry seasons of relatively equal duration). Most of the region's annual rainfall is experienced during the wet season, and very little precipitation falls during the dry season [5, 6].

The tropical climatic conditions are constituted by rains concentrated in November and March and a dry period that goes from April to October with haste in general inferior to 60 mm in the dry periods. In the large area of central Brazil [5], Aw climate is markedly seasonal, with strong longitudinal gradient (east-west) of annual rainfall from 1,300 to 1,900 mm and an opposite gradient (west-east) in the rainfall seasonality.

The conditions found in the regions of wet tropical climate produce, in great part, peculiarities of the Brazilian grounds (incident, constitution, formation, properties, rates, and environmental conditions), which are different from the considered ones in climate regions seasoned for which there were developed the systems of classification of traditional grounds [7].

2.1 Lateritic soils: definition and origin

The term laterita was used initially by Francis Buchanan in 1807 when, in travel to the west of India, he identified the use of a reddish ground that after drying was used like bricks in constructions of several sizes; the term laterita, however, includes a scale

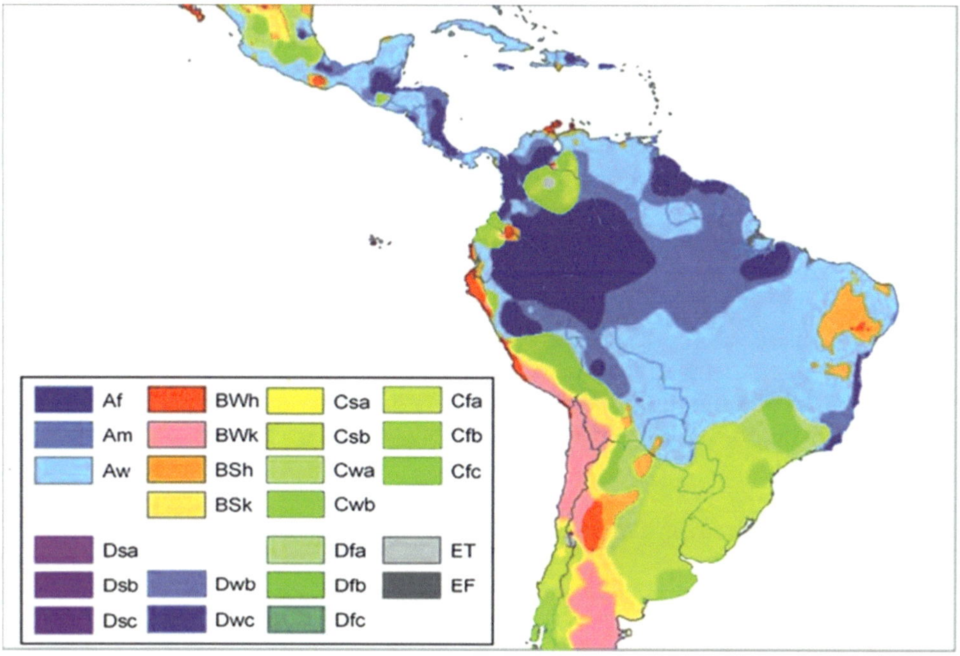

Figure 1.
Climate map of South America [4].

bigger of materials and of behavior varied [8]. However, it is known that the laterita was already used like building materials before that and his importance attached for the production of foods and of construction in function of the vast area of incident in the world, they are studied by several authors around the world as in Brazil, Africa, India, Australia, and other places [9].

The laterização is a process that makes part of the evolution of the relief [10] and [11] in which it takes place to lixiviation of alkaline ones, magnesium, and partially of the quartz and the consequent layers, the formation of lateritas, what are the mixture of hydroxides of iron and aluminum in varying proportions plus add up titania and other residue left. They can constitute micro-collected or collected cement of few centimeters of diameter in the womb of the soil [12].

Depending on the degree of laterization, the materials can be presented under several forms of texturais what go from not consolidated soft clays that can be broken under pressure of the fingers up to materials having enough edurecidos. That led to the use in the concepts literature empiricos of degree of hardness as "hard" or "soft" [8, 13]. However, these expressions guard little relation with mechanical properties of interest of the engineering. Since the variety of lateritas and the changes in his conditions due to environmental factors, his agreement to classifications that use purely morphological concepts, will not always be possible [14].

Another aspect of the formation of these grounds is that the lixiviation of the bases and of the sílica, nevertheless, can be incomplete and the distinction between two types of grounds is difficult to be done; in spite of the properties of two types of grounds, it is similar in terms of properties for the engineering [15].

As for the time of formation of the grounds, lateríticos appreciate that takes place in nearly 104 years, but there are evidences of which this formation is quicker in rocks with less content of quartz like basaltos in granites or rich sediments in quartz [16, 17].

Figure 2 illustrates the process of formation of the tropical grounds and the denominations used for the same.

As noted, the materials classified as lateritic owe their mechanical and hydraulic behavior to this process of "laterization" that promotes the leaching of basic cations and concentration of iron oxides and aluminum and additionally the predominance of clay minerals of the group of caulinites, low CTC. Studies have shown that soils formed under similar conditions tend to exhibit similar indices and engineering properties [19].

It is noted that Pedology, science that originated in the countries of the northern hemisphere, where soil formation processes are delayed due to cold winter or dry summers, meets challenges for soil description and classification in the tropics, including the difficulties of distinguishing soil from source material and the different horizons resulting from the intense pedogenetic processes of the tropical climate [10].

3. Characterization and properties of lateritic tropical soils

Lateritic soils (later Latin: brick) are shallow soils, typical of well-drained regions of the tropical wet regions, and have peculiar characteristics associated with the laterization process being the most important from the technological point of view,

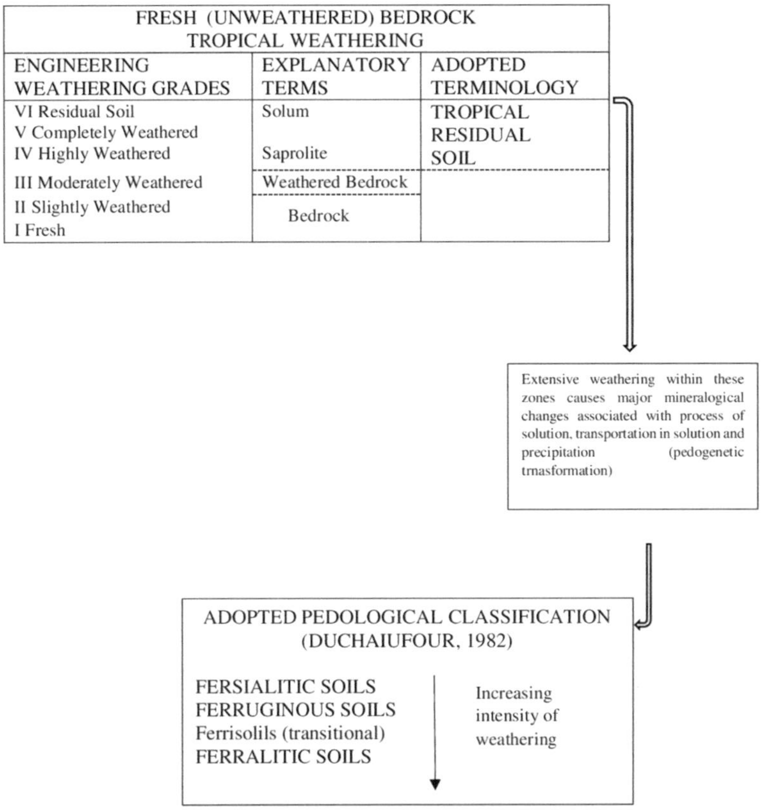

Figure 2.
Terms used in the description of tropical soils [18].

soil enrichment of iron- and/or aluminum-hydrated oxides, and the permanence of kaolinite as predominant and almost always exclusive clay. These minerals give the soils a typical color: red, yellow, brown, and orange [7, 15].

The saprolitic soils (sapro, Greek: rotten) are those resulting from the decomposition and/or *in situ* breakdown of the matrix rock by the action of the timeless agents and that maintains the structure of the source rock. They can be considered genuinely residual, because the particles that make it up remain in the same place as the matrix rock and can be called young residual soils, in contrast to the lateritic surface mature soils [7].

Saprolitic soils form layers underlying the layer of lateritic surface soil (or possibly other soil) generally appearing on the surface of the soil due to soil erosion or excavation due to man-made works. These soils are more heterogeneous and consist of a complex mineralogy containing minerals still in decomposition phase.

The characterization and evaluation of the geotechnical properties of residual soils are a complex subject, and there is the need for studies on their peculiar behavior for different purposes such as foundation, roads, stability of taludes, construction of earthworks, among others.

Figure 3 illustrates some variations that can be found in the intemperism profile of tropical soils and that contribute to the complexity of the approach, and **Figure 4** shows concretions formed by this material.

Field investigations of residual soils often relate to heterogeneous soil profiles vertically and horizontally, great structural complexity, and expected metastability due to the process of leaching and chemical decomposition, the presence of rock blocks immersed in matrix, among other aspects [21].

Despite the difficulties of naming these soils, there is a relative consensus that their characterization is made by conventional criterion, which is chemical, that is, would be lateritic soils all those in which the silica/sesquioxides ratio is greater than or equal to 2, and it is deeply weathered soil [22].

Despite the conventional definition, it is found in many situations that the behavior of these soils cannot necessarily be addressed by the conventional geotechnical project due to one or more of the following reasons [21]:

1. Soil state is variable due to complex geological conditions.

2. Classical constitutive models do not offer an approximation of their true nature.

3. These formations are difficult for sample and the soil structure cannot be reproduced in the laboratory. As a consequence, mechanical behavior and geotechnical properties should be evaluated directly from *in situ* test data for most geotechnical design problems.

4. There is also a limited experience collected and reported and the finding that parameter values are outside the most commonly found ranges for sand and clay formations of sedimentary soils.

5. Deposits are often unsaturated and the role of matrix suction and its effect on soil permeability and shear resistance must be recognized and accounted for.

These difficulties of lifting and characterization are detached also by authors [23–25], and others, according to which the tropical grounds it has the reputation of there are "problematic soils" because of without being fitted in the classification systems usually used as they were developed for temperate climates; there is also the need to use adequate methods for them, since the destruction of the cement and its original structure compromises the analysis of its behavior.

Variability in its engineering properties implies, in many situations, a difficulty to meet traditional specifications or consecrated use. An example put forward by [8] is that these materials usually have gaps in the graduation curve (e.g., in the coarse sand fraction); high plasticity indices (PIs 15-20) and CBR values below the minimum 80% are normally specified. An interesting discussion about unconventional materials and their research can be seen in [26].

The geotechnical behavior of these soils is therefore influenced more by their unsaturated condition and by factors such as their structure, macro- and microporosity, anisotropy, and genesis than by their stress history [27–29].

3.1 The lateritic soils in Brazil

As mentioned above, studies on tropical soils often exhibit a higher degree of difficulty because of their mineralogical, textural, and structural variability, and this is not an exception in the Brazilian territory. Several researchers have studied the behavior of these soils both in the context of the execution of works and for experimental purposes, in universities and research centers. One aspect to highlight is the contribution of foreign companies and professionals in recent years to enriching the knowledge and discussion of engineering problems from the exchange of ideas [30].

Although the country has soil surveys developed by EMBRAPA (Brazilian Agribusiness Research Company) and other research bodies, these are mainly for use in agriculture, without the geotechnical focus [31].

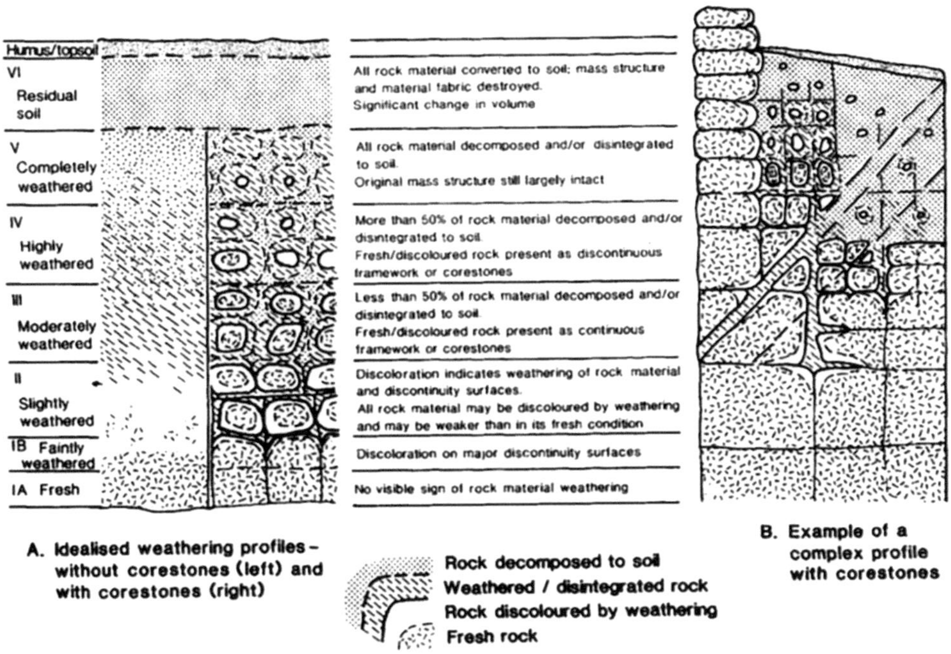

Figure 3.
Alteration profiles of tropical soils [16].

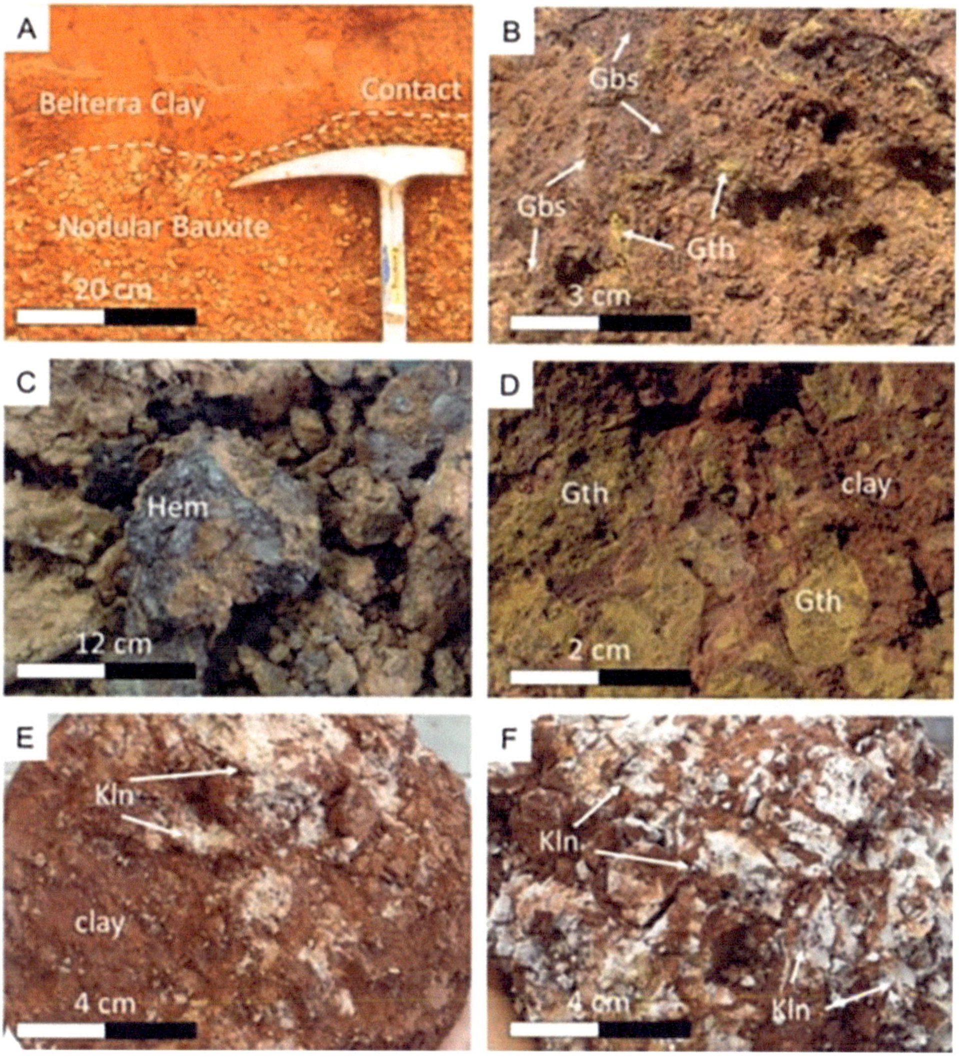

Figure 4.
Distinct mesoscopic aspects of the lateritic materials at the Rondon do Pará bauxite deposit. A. Contact between Belterra clay and Nodular Bauxite; B. massive bauxite; C. fragments of iron crust with hematite; D. ferruginous bauxite with oxyhydroxide clasts in a gibbsite matrix, strong goethitized; E: massive bauxite base with kaolinite; F. mottled zone in the basal clay. Approximate scale, drillcore HQ diameter of 9.65 cm (3.5"). Gbs-Gibbisite, Gth-Goethite, Hem-Hematite, Kln-Kaolinite [20].

In this context, the peculiarities of Brazilian soils (occurrence, constitution, formation, properties, indices, and environmental conditions) are therefore different from the conditions found in the temperate climate regions where the traditional soil classification systems were developed [4, 7].

It is thus observed that the physical, chemical, biological, pedological, and geomorphological processes vary throughout the area of occurrence of these soils and also in Brazil, where they are distributed over 80% of the territory, as shown in **Figure 5** [32].

Since knowledge of where a work will be deployed depends primarily on well-designed and developed local research, one of the important aspects for the development of an assertive engineering project is the description of the soil profile. Expedited forecasts of collapsible or expansive soil behavior could be inferred from pedological classifications, where in addition to soil identification and classification, information about soil genesis is provided. This is because there is a close dependence on the tropical humid climate of the changing soils in relation to the matrix rock, as, for example, granites decompose originating mycaceous soils with particles of clay and sand, and basalts change basically in clays [3].

Of course, the use of generic profiles is inadvisable and local research can in no way be replaced. Consequently, the use of the geological description of the soil profile in engineering projects is considered essential, and the ignorance of the soil profile leads the designer to make predictions with a degree of uncertainty above that tolerated in the standards. On the other side, when the origin was known and the characteristics of the whole region and of the profile in an individual place the

Figure 5.
Area of occurrence of laterite soils in Brazil represented by dark brown colors and hachuras [32].

foresight becomes more assertive, reducing risks, costs, and creating solutions more appropriate to each situation [33, 34].

The occurrence of porous layers of clay or sandy texture and materials with varying degree of intemperization is frequent in Brazilian soil profiles, requiring in some situations, analyses and more complex models regarding the geotechnical behavior of the soil. **Figures 6** through **8** present some profiles of lateritic soils found in engineering works in Brazil.

In relation to the research methods, one of the most used in Brazilian geotechnical engineering for underground research is the survey of simple recognition of soil with SPT test as highlighted [36, 37], its execution currently governed by the Brazilian

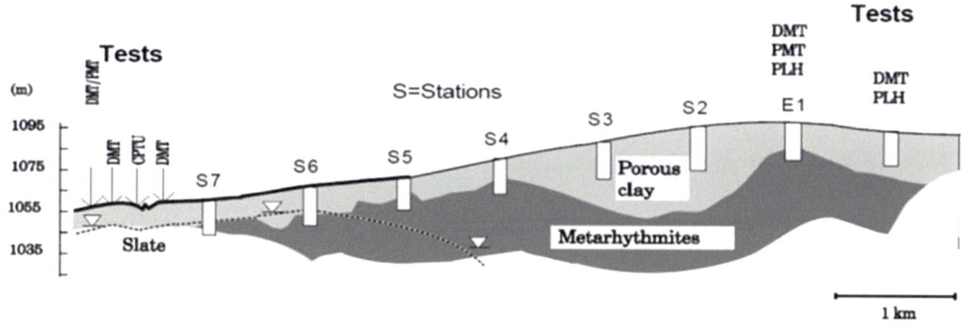

Figure 6.
Soil profile along the tunnel, South Wing, Brasilia [4].

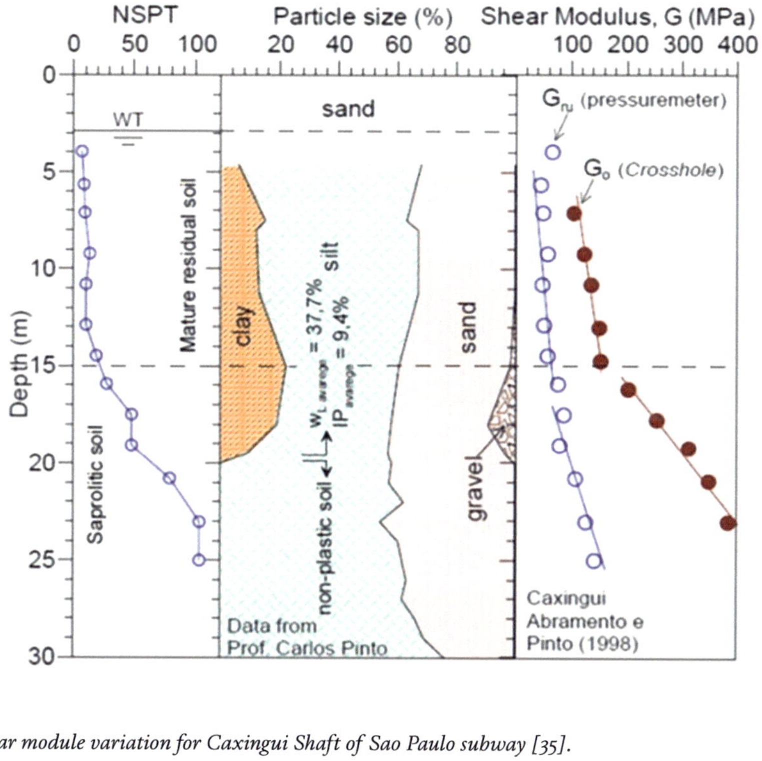

Figure 7.
Profile of shear module variation for Caxingui Shaft of Sao Paulo subway [35].

NBR 6484:2020. Despite the existence of other methods, the tendency to use this remains, either for its lower cost compared with the other methods (CPT, DMT, geophysical, or other) or for the fact that it does not require specialized labor, for the empirical correlations that are established from or by which other methods are little publicized in the universities.

This tendency of using the sounding SPT can be observed in an abreve consultation to the site of the ABMS (Brazilian Association of Mechanics of Grounds) in which nearly 100 articles are listed that wrap different applications of the method in several types of work (https://philos.sophia.com.br/terminal/8530/) as in [32, 38–43] and many other authors.

However, in some situations where the size of the work, geological complexity, or even academic studies so require, and other methods are used such as CPT (cone penetration test) or DMT (Dilatometer Marchetti Test) as well as the collection of undeformed samples for laboratório testing (triaxial compression, shear strength, deformation, modulus, and others) as exemplified by the work of [22, 30, 44–55].

3.2 Geotechnical properties of lateritic soils in some sites in Brazil

As previously mentioned, lateritic soils can present quite varied behaviors, which requires the implementation of projects that take into account their geotechnical properties, obtained from field and/or laboratory tests. The growing urbanization and verticalization of cities in the center-south region of Brazil, as well as the increase in infrastructure works, lead to increasing challenges in foundation design, since the foundation elements must consider the high loads to be distributed frequently. For this, the use of piles has been the most common foundation option, since in some situations, geotechnical limitations occur due to soil properties such as high porosity and/or the collapsible character or excessive settlements in the face of loads, which do not allow the use of direct foundation.

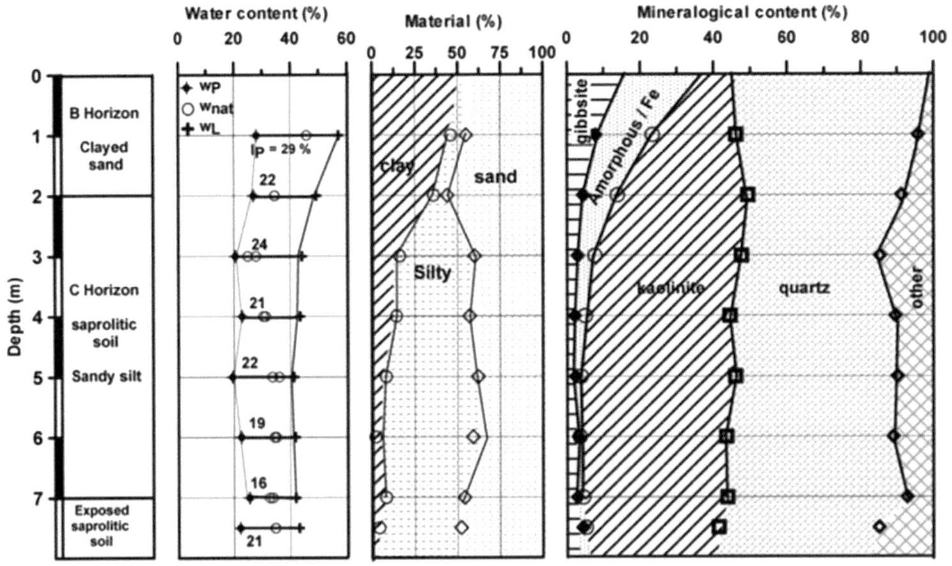

Figure 8.
Site characteristic profiles [15].

One example of this behavior is the soils of the city of Brasilia, located in the state of Goiás, which is located in an area of highly weathered tropical soil, with high levels of aluminum and iron. As a consequence of the high porosity of the cemented structure of this soil, known as "porous clay," it presents great structural instability and can sorer that is highly unstable and can suffer changes in volume (collapse) due to changes in saturation and stress state. The possibility of sharp deformations must be considered since this type of soil covers 80% of the area of the municipality [53–55].

Figure 9 illustrates the SPT index strengths in the study conducted by [53, 54] and the strength parameters of unsaturated porous clay, according to Mohr-Coulomb criteria that can be considered as Cohesion angle ranging from 20 to 34 KPa, friction angle () between 25 and 28° and Young's modulus varying 1–8 Mpa, Coefficient of collapsibility is 0–12%, Coefficient of permeability is 10^{-06}–10^{-03} cm/s [55, 56]. The tests to evaluate the granulometry of the soil composed of sieving and sedimentation, in addition to the Atteberg limits test, allowed the classification of the soil as CH by the Unified Classification system, with a plastic Index of 12% and a natural unit weight around 15 kN/m^3. This porous clay layer has a variable thickness of 20–30 m and NSPT indexes between 2 and 3 strokes with a deep water level, and in some cases reaching a depth of 5 m. Liquid limit LL = 50–80%, plastic limit PL = 35–50%, and water content w = 35–55%. The clay fraction, that is, the percentage of soil particles less than 2 μm lies between 70 and 55%. The percentage of fines (less than 60 μm in diameter,) varies from 70 and 80% [55].

Considering the characteristics observed in the drilling and laboratory tests, the authors state that a foundation option that has been used in the city of Brasilia and neighboring cities is the Alluvial Pile Anker, a new type of small diameter foundation characterized by fast execution, with technical and economic advantages over precast piles. It consists of drilling small diameter piles, where a 2 ½″ tube, 50 cm longer than the depth of the hole, with a cutting tip (**Figure 10**), is installed in the ground at very high speed, and the soil is drilled through rotation. The hole is filled with cement, and after it has been drilled, the capping is made with precast-reinforced concrete or steel sheeting on each pile. A gravel backfill is placed between the capstones at the same height, and a geogrid is placed over it, followed by a transition backfill that acts as a stress dissipator [53, 57].

It should be noted that in foundations embedded in lateritic or collapsible soils, the rigor in the design process and in the design should be greater, because the behavior of these foundations often differs from the classical models adopted and presented in the technical-scientific literature, being possible to observe a nonlinearity of the soil behavior due to variations in the soil parameters that control its behavior: modulus of deformation and shear modulus of the soil [58]. It is fundamental for pile foundation design like aspects such as the relative stiffness of the lateritic soil of the first layer when not saturated; the collapsibility of this soil, especially if the piles are totally embedded in this layer and the evaluation of the ultimate strength, due to its own executive process, be taken into consideration.

Studies conducted in collapsible lateritic soils in the city of Campinas, state of São Paulo, where most of the foundations employed are deep, with auger piles being the most commonly used, show the use of foundations executed as staked foundations, for example, a foundation element where the piles under the radier are interrelated may have greater efficiency in the reduction of settlements, because the greater contact of the surface foundation element contributes to the performance of load capacity and settlement reduction for the system [59, 60].

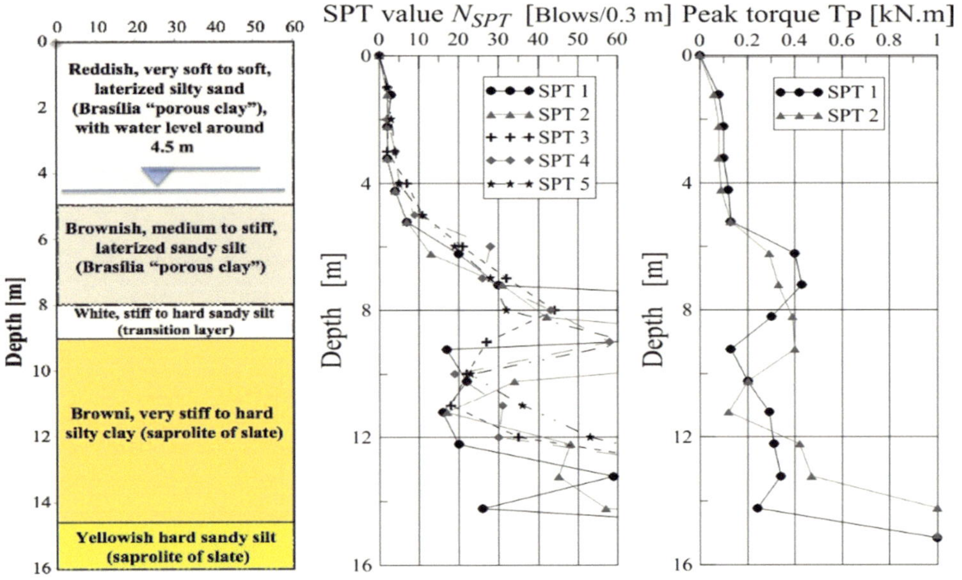

Figure 9.
Stratigraphy and SPT-SPTT results [53].

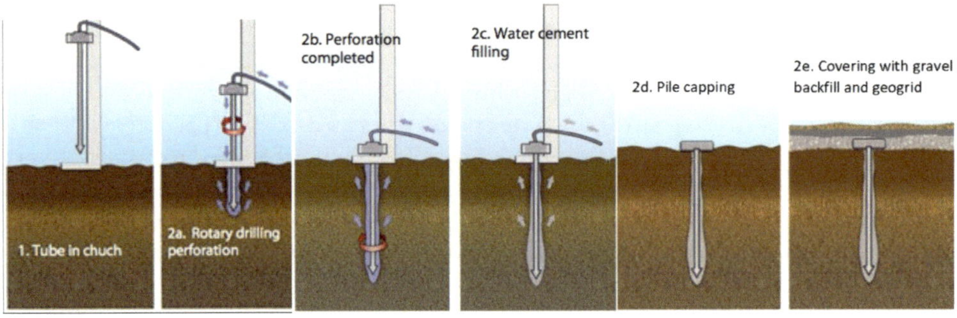

Figure 10.
Alluvial Anker pile construction process [57].

In the city of São Paulo, the construction of an extensive subway network allowed obtaining geotechnical parameters of soils existing in the São Paulo Basin from the study of 12 different sites, which demonstrated the heterogeneity of the profiles, comprising alternating layers of sandy clays and clayey sand with silt fractions. However, the horizontal stress index (Kd) revealed values greater than 2, confirming the overconsolidation of the variegated soils, which had been previously reported in the literature [22]. The *in situ* tests allowed obtaining information about the stress history of the sampled soils and despite the variability of the soils, authors point out that the corrections obtained are consistent with results presented in the technical literature, confirming the potentiality of the piezocone and dilatometer (DMT) tests [33].

It is noteworthy that lateritic soils can also present variation in their behavior as a function of matrix suction variation, and therefore, their geotechnical investigation should be careful [21, 26, 33, 58, 61, 62].

Furthermore, one must consider the resistance variation presented by laterite soils that vary considerably with depth, according to the influence of such factors as parent rock, depth of the water table, topography, degrees of decomposition, laterization, and desiccation, as well as mineralogical composition. Also in relation to mineralogy, if the clay present in the soil has the presence of iron oxide in ferric state, the soil is essentially stable and no changes are expected and therefore, standard tests can be employed for soil characterization [8, 32, 63].

4. Final considerations

Tropical soils occur over a large area of the planet, occupying about 40% of the surface. The tropical climate is responsible for the laterization process, which generates well-drained soils, porous, reddish in color, and with characteristics different from those of temperate soils and that may require, by their nature, solutions different from those proposed by classical soil mechanics.

Despite the importance of these soils, there is still no integrated database on their characteristics and behavior. In Brazil, residual and saprolitic soils are a challenge to engineering because they range from highly weathered and well-drained, porous soils in tropical and subtropical climates to thin and poorly developed soils in regions of the country where the drier climate predominates.

This chapter has aimed to approach some aspects of the soils in Brazil, without the pretension of exhausting the subject, which is of great interest to the country and others that are located in the tropical region.

There is a need for further investigation of soils not only in terms of fertility or application to agriculture, as is often the case, but also in terms of geotechnical aspects for the execution of foundations and roads, retaining structures, among others, so that field and laboratory tests can be conducted, not only in the south and southeast regions, but also in other regions, considering the constant expansion of the country's infrastructure, including international partnerships.

Author details

Maisa Comar Pinhotti Aguiar
Universidade do Estado de Minas Gerais (UEMG), Unidade de João Monlevade (MG), Brazil

*Address all correspondence to: maisa_trabalhos@outlook.com

References

[1] Ishak MF, Zolkepli MF, Affendy M. Tropical Residual Soil Properties on Slopes. International Journal of Engineering Technology and Sciences (IJETS). 2017;**8**(1). DOI: 10.15282/ijets.8.2017.1.12.1087

[2] Wesley LD. Fundamentals of Soil Mechanics for Sedimentary and Residual Soils. John Wiley & Sons, Inc; 2009 ISBN: 978-0-470-37626-3

[3] Massad F. Escavações a céu aberto em solos tropicais: Região Centro-Sul do Brasil. São Paulo: Oficina de Textos; 2005 ISBN 85-86238-39-2

[4] Ortigão JAR. Dilatometer tests in Brasília porous clay. In: Proc. 7th Intern. Congr. Intern. As. Eng. Geol, Lisboa, Portugal. 1994

[5] Alcarde AC et al. Koppen's climate classification map for Brazil. Meteorologische Zeitschrift. 2014;**22**(6):711-728. Available from: https://www.schweizerbart.de/papers/metz/detail/22/82078/Koppen_s_climate_classification_map_for_Brazil (published online January 2014)

[6] Schmidt A, Eitzinger A, Sonder K, Sain G. Tortillas on the Roaster (ToR) Central American Maize-Bean Systems and the changing Climate: Full Technical Report. Baltimore: Maryland, USA; Cali: Colombia. Mexico, D.F.: Mexico; 2012. p. 127

[7] Nogami JS, Vilibor DF. Pavimentos de baixo custo para vias urbanas. São Paulo: Arte e Ciência; 2009 193 p

[8] Maignien, R. Review of Research on Laterites. 1966. Natural Resources Research IV, UNESCO, Paris, 148 pp. Available from: https://unesdoc.unesco.org/ark:/48223/pf0000071101, acess in 14/12/21

[9] Aleva AJJ. The Corlat Handbook: Guidelines and Background Information for the Description of Laterite Profiles for Interdisciplinar Use. Wageningen, Netherlands: Corlat Technical Information n. 1. ISRI-Information Soil Reference and Information Centre; 1993

[10] Jenny H. Factors of soil formation: A System of Quantitative Pedology. New York: Dover Publications, Inc.; 1994

[11] McFarlane MJ. Mechanisms for lateritisation and the formation of erosion surfaces in parts of east and southern Africa. Bulletin de la Société géographique de Liège. 1991;**27**(1991):149-155

[12] Volkoff B. Red and lateritic soils: World scenario (6th chapter). In: Red & Lateritic Soils. Vol. 1. Rotterdam, Nehterlands: Balkema; 1998

[13] Nettergerg N, (org). Review of Specifications for the Use of Laterite in Road Pavements. Africa Community Access Programme (AFCAP). Available from: https://assets.publishing.service.gov.uk/media/57a089e040f0b652dd000450/Laterite-Final-AFCAPgen124-v140525.pdf; 2014 [Accessed: 18 December 2021]

[14] AID. Agency for international Development. Laterite e lateritic soils and Other problems soils in Africa. Baltimore: USA: Lion Associates; 1971 Available from: https://pdf.usaid.gov/pdf_docs/Pnaah867.pdf

[15] Futai MM, Almeida MSS. An experimental investigation of the mechanical behaviour of anunsaturated

gneiss residual soil. Geotechnique. 2005;**55**(3):201-213

[16] Vaughan PR. Tropical Residual Soils Geological Society Engineering Group Working Party Report. Quarterly Journal of Engineering Geology and Hydrogeology (1990). 1990;**23**(1):4-101. http://dx.doi.org/10.1144/GSL.QJEG.1990.023.001.01

[17] Vaughan PR, Maccarini M, Mokhtar SM. Indexing the engineering properties of residual soil. Quarterly Journal of Engineering Geology and Hydro-geology. 1988;**21**(1):69-84

[18] Fookes PG. A review: Genesis and classification for tropical residual soils for engineers. Balkema: Rotterdam, ISBN 90 5410 007 9: Geotechnics in African Environment; 1994. Available from: https://edisciplinas.usp.br/pluginfile.php/5362924/mod_resource/content/1/Tropical%20Soil_Genese_Classif.pdf

[19] Prass AS et al. Estudo comparativo de diferentes sistemas de classificações geotécnicas aplicadas a solos do município de Alta Floresta – MT. In: Anais 33° Congr. Pesquisa e Ensino em Transporte da ANPET, Camboriu, SC. 2019 Available from: https://www.anpet.org.br/anais/documentos/2019/Infraestrutura/Solos%20e%20Concreto%20I/6_695_AC.pdf

[20] de Oliveira SB, da Costa ML, Prazeres Filho HJ. The Lateritic Bauxite Deposit of Rondon do Pará: A New Giant Deposit in the Amazon Region. Economic Geology· Northern Brazil. 2016. DOI: 10.2113/econgeo.111.5.1277

[21] Schnaid F, Huat B. Sampling and testing of tropical residual soils. In: Huat BBK, Toll DG, Prasad A, editors. Handbook of Tropical Residual Soils Engineering. Boca Raton: CRC Press; 2012. p. 24 Routledge Handbooks Online

[22] Caldo M, Massad F. Geotechnical parameters for the variegated soils of São Paulo formation by means of *in situ* tests. Soils and Rocks, São Paulo. 2016;**39**(2):189-200

[23] Brand EW. Predicting the performance of residual soil slopes. In: Proceeding of 11th International Conference on Soil Mechanics and Foundation Engineering, San Francisco, August 12-16. 1985

[24] Toll DG. Tropical soils. In: ICE Manual of Geotechnical Engineering, chapter 30. 2012. DOI: 10.1680/moge.57074.0341

[25] Tuncer ER. Engineering behavior and classification of lateritic soils in relation to soil genesis [Retrospective Theses and Dissertations]. 1976. 5712. https://lib.dr.iastate.edu/rtd/5712

[26] Schnaid F, Lehane BM, Fahey M. *In situ* test characterisation of unusual geomaterials. In: Proc. 2nd Int. Conf. on Site Charact. Vol. 1. Porto: Milpress; 2004. pp. 49-74

[27] Rocha BP. Geotechnical characterization of unsatureted tropical soil by in situ testes. São Carlos [Phd Thesis]. Escola de Engenharia de São Carlos, Universidade de São Paulo (in portuguese); 2018 242 p

[28] Vargas M. Engineering Properties of Residual Soils from Southern-Central Region of Brazil. Proc. 2nd Int.Cong. Vol. 1. SãoPaulo: IAEG; 1974. pp. 5.1-5.26

[29] Vargas M. Structurally unstable soils in Southern Brazil. Conference, Session four. 11F, 2T., 4R. Proc 8th Int. Conf on Soil Mech Found. Eng, Moscow. Vol. 21973. pp. 239-246

[30] Alburquerque PJR, Rodrigues TG. Assessment of Results of CPT Tests in

Porous Lateritic Unsaturated Soil from Campinas, Brazil. In: Fifth International Conference on Geotechnical and Geophysical Site Characterization (ISC'5). Theme 10 - Characterisation in Rock and Residual Soil. 2016 Available from: https://www.issmge.org/publications/author/t-g-rodriguez

[31] Cooper M et al. A National Soil Profile Database for Brazil Available to International Scientists. Soil Sci. Soc. Am. J. 2005;**69**:649-652. DOI: 10.2136/sssaj2004.0140

[32] de Araujo CBC, Dantas Neto SA. Caracterização de Solos Lateríticos para Utilização em Pavimentos de Baixo Custo na Cidade de Canindé/CE. In: XII Congresso Brasileiro de Mecânica de Solos e Engenharia Geotécnica. Goiânia/GO, Brasil, september/2014. 2014

[33] de Carvalho JC et al. Solos não 0Paulo: Associação Brasileira de Mecânica dos Solos e Engenharia Geotécnica (ABMS); 2015. p. 805

[34] Gusmão Filho JA. Fundações. Recife: Editora Universitária UFPE; 1998 1998 - 345 páginas

[35] Futai MM, Cecilio MO Jr, Abramento M. Shear strength and deformability of residual soils from the Sao Paulo metropolitan area. In: Twin Cities - Soils from Sao Paulo and Curitiba Metropolitan Areas, Chapter: 7. 2012 Available from: https://www.researchgate.net/publication/323103604_Shear_Strength_and_Deformability_of_Residual_Soils_from_the_Sao_Paulo_Metropolitan_Area

[36] Menezes SM, Sobrinho MD. Evaluation of SPT-test- energies, torques, shocks, velocities and limitations of your measurements. Proc. X Brazilian Conference on Soil Mechanics and Foundation Engineering. 2010;**2**:369-366

[37] Milititsky J, Consoli N. Foundation engineering- Brazilian practice. Proceedings 8th Brazilian Conference on Soil Mechanics and Foundation Engineer ing. Porto Alegre: Brazil; 1986

[38] de Alcantara Junior AP, Conciani W, Vidrih C. Aplicabilidade do SPT no dimensionamento de estacas apiloadas em solos colapsíveis. In: Anais XII Congresso Brasileiro de Mecânica dos Solos e Engenharia Geotécnica (COBRAMSEG), 2002. São Paulo: ABMS; 2002

[39] Almeida MSS, Marques MES, Fonseca OA. Características do solo de fundação da terceira pista do aeroporto de Guarulhos. In: Anais XII Congresso Brasileiro de Mecânica dos Solos e Engenharia Geotécnica (COBRAMSEG), 2002. São Paulo: ABMS; 2002

[40] Alves PC, Souza NM. Cartografia Geotécnica para Obras Subterrâneas no Plano Piloto de Brasília. In: COBRAMSEG 2010: COBRAMSEG 2010: Engenharia Geotecnia para o desenvolvimento, inovação e sustentabilidade. ABMS; 2010

[41] Coutinho RQ, Schnaid F. CPT Regional Report for South America. 2nd International Symposium on Cone Penetration Testing, Huntington Beach, CA, USA, May 2010; 2010. Available from: http://www.emtermos.com.br/abms/site/conteudo/CPT2010RR9SAmerica.pdf

[42] Gonçalves HHS, de Oliveira NJ. Parâmetros geotécnicos das argilas de Santos. In: XII COBRAMSEG 2002: anais. São Paulo: ABMS; 2002

[43] Probst CA et al. Análise Comparativa de Métodos de Determinação da Capacidade de Carga em Estacas Hélice Contínua com Ensaios de Prova de Carga Estática Realizados em Uberaba-MG.

In: XIX Congresso Brasileiro de Mecânica dos Solos e Engenharia Geotécnica Geotecnia e Desenvolvimento Urbano COBRAMSEG 2018 – 28 de agosto a 01 de setembro, Salvador, Bahia, Brasil. ABMS; 2018

[44] Coutinho RQ, de Castro BPF, Dourado KCA. Análise de fundação profunda em solo colapsível da Penitenciária Regional de Eunápolis – BA. COBRAMSEG. Engenharia Geotecnia para o desenvolvimento, inovação e sustentabilidade. ABMS; 2010, 2010

[45] Giacheti HL et al. O ensaio de piezocone sísmico: execução, interpretação e exemplos de aplicação. In: Conference: SEFE 7: Seminário de Engenharia de Fundações Especiais e Geotecnia, São Paulo, SP. 2012 Available from: https://www.researchgate.net/publication/327750954_O_ensaio_de_piezocone_sismico_execucao_interpretacao_e_exemplos_de_aplicacao#fullTextFileContent

[46] Giacheti HL, De Mio G, Peixoto ASP. Cross-hole and seismic CPT tests in a tropical soil site. GeoCongress 2006. Vol. 2006. Geotechnical Engineering in the Information Technology Age; 2006. p. 92 Available from: http://hdl.handle.net/11449/69432

[47] Goulart RL, Angelim RR, Cunha RP. Ensaios Dilatométricos em Aterro Compactado da Barragem do Ribeirão João Leite. Conference XIX Congresso Brasileiro de Mecânica dos Solos e Engenharia Geotécnica Geotecnia e Desenvolvimento Urbano (COBRAMSEG) 2018, Salvador, BA. 2018. Available from: https://www.researchgate.net/publication/349773968_Ensaios_Dilatometricos_em_Aterro_Compactado_da_Barragem_do_Ribeirao_Joao_Leite [Accessed: 12 December 2021]

[48] Mantaras FM, Odebrechet E, Schnaid F. Using piezone dissipation test to estimate the undrained shear strenght in cohesive soil. Can Geotech J. 2015;**52**:318-325, dx.doi.org. DOI: 10.1139/cgj-2014-0176

[49] Pontes CV et al. Geological-geotechnical risk mapping of gravitational mass movements in an urban area in Colombo, Brazil. 2021. Available in: https://www.scielo.br/j/soiroc/a/L4pXRpLqLWr4LxJXW9FHdQN/. DOI: 10.28927/SR.2021.070721

[50] Rocha BP, Castro BAC, Giacheti HL. Seismic DMT Test in a Non-Text Book Type Geomateria. 2015. Available from: https://www.marchetti-dmt.it/conference/dmt15/papers%20DMT%202015%20(pdf)/128.pdf

[51] dos Santos MD, Bicalho KV. Proposals of SPT-CPT and DPL-CPT correlations for sandy soils in Brazil. Journal ofRock Mechanics and Geotechnical Engineering. 2017 Available from: https://www.researchgate.net/publication/319442619_SPT-CPT_and_DPL-PT_correlations_proposals_for_sandy_soils_in_Brazil

[52] Soares CSM et al. CPT in a tropical collapsible soil (chapter). In: Hicks, Pisanò, Peuchen, editors. Cone Penetration Testing 2018. The Netherlands, ISBN 978-1-138-58449-5: Delft University of Technology; 2018 Available from: https://www.taylorfrancis.com/chapters/oa-edit/10.1201/9780429505980-82/cpt-tropical-collapsible-soil-soares-danziger-jannuzzi-martins-andrade

[53] Mendoza C, Cunha R, Ruge JC, Lizcano A. Some contributions to the behavior of piled-rafts made up by self-drilling piles founded in a tropical soil of Brazil. Electronic Journal of Geotechnical Engineering (EJGE);**19**(Z5):17181-17194

[54] Mendoza CC, Cunha R, Lizcano A. Mechanical and numerical behavior of groups of screw (Type) piles founded in a tropical soil of the Midwestern Brazil. Computer and Geotechnics. 2015;**67m**:187-203 Available from: https://www.sciencedirect.com/science/article/abs/pii/S0266352X15000439

[55] Ortigao JAR, Macedo P. Large settlements due to tunneling in porous clay, presented at the International Tunneling Conference, Toulon, Oct. Paris: Tunnels et Ouvrages Souterrains, no. 119, Sept-Oct 93; 1993. pp. 245-250

[56] Jacazz S, de Matos YMP, Monteiro FF, Cunha RP, Ruges JC, Gassl G. Evaluation of analytical and numerical techniques to simulate Curtain Pile walls in a tropical soil of the federal district of Brazil, Geotechnical Engineering. Journal of the SEAGS & AGSSEA. 2020; 512 ISSN 0046-5828

[57] Barbosa M. Alluvial anker como alternativa para fundações em argila mole (In Portuguese). Tech. rep. Brasilia, Brazil: Solotrat Engenharia Geotecnica Ltda.; 2009

[58] Garcia JR, Albuquerque PJR. Model of nonlinear behavior applied to prediction of settlement in deep foundations. DYNA. 2018;**85**(205):171-178. DOI: 10.15446/dyna.v85n205.68523

[59] Garcia JR. Experimental and numerical analysis of piled rafts executed in soil of the Campinas / SP region [thesis (Doctorate)]. School of Civil Engineering, Architecture and Urbanism, State University of Campinas (UNICAMP); 2015, 2015 (in portuguese)

[60] Scallet MM. Behavior of bored piles of small diameter in collapsible and lateritic soil in the region of Campinas/SP [Thesis (Master of Science)]. School of Civil Engineering,Architecture and Urbanism, State University of Campinas (UNICAMP) 167 p; 2011 2011 (in portuguese)

[61] Costa YD, Cintra JC, Zornberg JG. Influence of matric suction on the results of plate load tests performed on a lateritic soil deposit. ASTM Geotechnical Testing Journal. 2003;**26**(2)211-227. Available from: https://www.caee.utexas.edu/prof/zornberg/pdfs/AJ/Costa_Cintra_Zornberg_2003.pdf

[62] Cunha RP, Silva CM, Matos YMP, Cárdenas JCR. Experiência de Projeto e Investigação de Cortinas de Estacas Justapostas no Distrito Federal. Conference: 4° Congreso Internacional de Fundaciones Profundas de Bolivia, Santa Cruz de la Sierra. 2019

[63] Gidigasu MD. Identification of problem laterit soils in Highway engineering: a review. In: 53rd Annual Meeting of the Highway Research Board: Columbia: USA, Transportation Research Record Issue Number: 497. Transportation Research Board. ISSN: 0361-1981; 1974 Available from: https://onlinepubs.trb.org/Onlinepubs/trr/1974/497/497-010.pdf

Chapter 2

Perspective Chapter: Interpretation of Deep Foundation Load Test Data

Jon Sinnreich

Abstract

Static load tests are seen by many practitioners as the best techniques to approximate in-service conditions for deep foundation elements and to validate analytical model predictions of capacity and settlement. Full-scale static load tests are fairly expensive to implement, especially as part of a pre-construction investigation when equipment and personnel must be mobilized to site separately to specifically install the test element(s). Test elements are often instrumented with strain gages to determine the load distribution during the test. Correct installation of gages and interpretation of the resulting data is critical to properly evaluate the test results and recoup the significant investment made in conducting the test. This paper discusses several key points in the interpretation of strain gage data in deep foundation load tests.

Keywords: strain gages, deep foundation load testing, tangent stiffness, incremental rigidity, t-z curves

1. Introduction

Static load tests are seen by many practitioners as the premier techniques to approximate in-service conditions for deep foundation elements and to validate analytical model predictions for load-bearing capacity and settlement. Full-scale static load tests are fairly expensive to implement, especially as part of a pre-construction investigation when equipment and personnel must be mobilized to site separately to specifically install the test element(s). Test elements are often instrumented with strain gages to determine the load distribution during the test. Correct installation of gages and interpretation of strain data is critical to properly evaluate the test results and recoup the significant investment made in conducting the test.

The ultimate product of a foundation test strain data analysis is often a set of curves which model the non-linear unit soil response to shear and bearing load (typically called 't-z' and 'q-z' curves, respectively). These curves are useful to model the foundation response to load [1]. Strain gage data is utilized to compute both the shear and bearing ('t') and displacement ('z') portions of the curves.

In the first section, statistical results collected by the author in two large-scale test programs involving multiple test foundations each are analyzed to investigate the optimal positioning of strain gages in a test element. In the second section, the conversion of strain to force via the rigidity function is discussed. In the third section, use of strain data to properly calculate zone displacement is derived.

2. Optimal strain gage arrangement

The discussion in this section was originally published in [2]. The assumption of axial plane-strain is significant to converting measured strain to axial force. Eccentric stress in the foundation element whether due to inclined or eccentric loading, uneven soil resistance, irregular foundation cross-section shape or other reasons will cause an uneven distribution of strain across the element cross-section. Based on Euler beam theory, the total strain is a superposition of axial strain (which is required to compute axial load) and bending strain which is disregarded. In axial compressive or tensile load testing of foundations it is presumed that axial strains due to applied loading will be significantly greater than incidental bending strains due to load eccentricity or other second-order causes. Total strain is assumed to be linearly distributed across the plane of the element, and the net average axial strain will intersect the centroid of the element. Therefore, obtaining the strain at the centroid is key to computing the net axial force.

The theoretical performance of gages arranged in various configurations and then averaged for the purpose of force calculation is validated using statistical results collected during the course of two large-scale test programs (the 'Florida' case history and the 'California/Nevada' case history, respectively), each involving multiple test foundations.

Normally, two or more strain gages are installed in a test foundation per level, attached to the steel reinforcement. Spacing the gages symmetrically around the perimeter allows for an estimate of the strain at the centroid to be computed as an average of the individual strain measurements. One opposed pair of gages is the typical arrangement. In the Florida case history test program, the owner specified three gages per level. It was not explicitly stated, but the implied arrangement was an equal spacing of 120° around the perimeter of the pile reinforcement cage (**Figure 1**).

Strain gages installed in cast-in-place foundation elements the field have a percentage mortality rate (probability of failure), designated λ. This is most often due to installation procedures for deep foundations. When constructing drilled shafts,

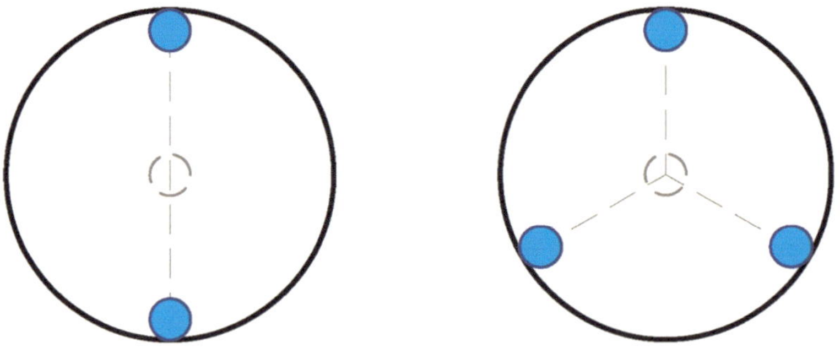

Figure 1.
Typical arrangement of opposed pair and triplet strain gages in pile cross section with computed average (dashed lines).

reinforcement cages must be lifted by crane, rotated from horizontal to vertical and then inserted into the shaft excavation. Concreting then takes place, either via the tremie method or by gravity pour, both of which are dynamic procedures with many opportunities to damage a gage. For auger-cast-in-place (ACIP) piles, the relatively slender and flexible reinforcement cage is lifted into the air at the head only for rapid insertion into fluid grout, which induces a 90° bend into the cage as it is lifted. In the Florida case history testing program, from a total of 657 sisterbar vibrating wire strain gages installed in eleven bi-directional ACIP test piles, seventeen gages failed to function during testing, yielding a mortality rate λ of 2.6%.

To estimate the strain at the centroid of the foundation element, the symmetrically-arranged gages at a given level are averaged. All the gages at a given level must function in order to compute the average at the centroid. Given k gages at a level, the probability of success S in this situation is computed as the simultaneous probability of survival of all the gages:

$$S_k = (1 - \lambda)^k \tag{1}$$

Even though in practice if a single gage fails the remaining gage(s) are often still utilized to estimate the average strain, this is not optimal since the resulting average is now off the centroid and therefore may not be representative of the average axial load if an uneven strain distribution is present in the cross-section due to bending stress.

To evaluate the potential significance of the difference between using an opposed-pair average and a single gage (assuming its opposite malfunctioned), data from a total of 207 pairs of functioning opposed gage pairs in the eleven axial test piles in the Florida case history is analyzed. A relative difference is computed for each logged reading of each opposed gage pair:

$$d = \frac{\left|\varepsilon_1 - \varepsilon_{avg}\right|}{\varepsilon_{avg}} \text{ where } \varepsilon_{avg} = \frac{\varepsilon_1 + \varepsilon_2}{2} \tag{2}$$

For each gage pair, the differences are averaged for all increments of loading. The resulting 207 data points are plotted on a histogram, and a log-normal probability distribution function is fitted to the resulting data (**Figure 2**).

The results of this analysis indicate that for this data set, the mean difference between data from a single gage and the average of the opposed pair is 15.3%, a significant dissimilarity. The inset figure plots the difference between individual and averaged strains as a percentage versus maximum average strain, which ranged from single digits of microstrain in gage levels near the ground surface to over 1000 microstrain in the vicinity of the bi-directional jacks. Although several of the highest individual difference values correspond to the smallest maximum strains, there is a fairly even distribution and no strong correlation to absolute values of strain, indicating the high mean difference is not confined to gage levels recording relatively little total strain (in other words, due essentially to a low signal-to-noise ratio). Obtaining a good measure of the average strain, rather than relying on an off-center result is thus crucial to computing the correct axial force.

Using Eq. (1), the surprising conclusion is reached that installing three equally-spaced gages per level (presumably for additional redundancy) actually results in a lower probability of successfully in obtaining the average strain at the pile centroid (92.4%) than by using two gages in an opposed pair (94.9%, using the numeric

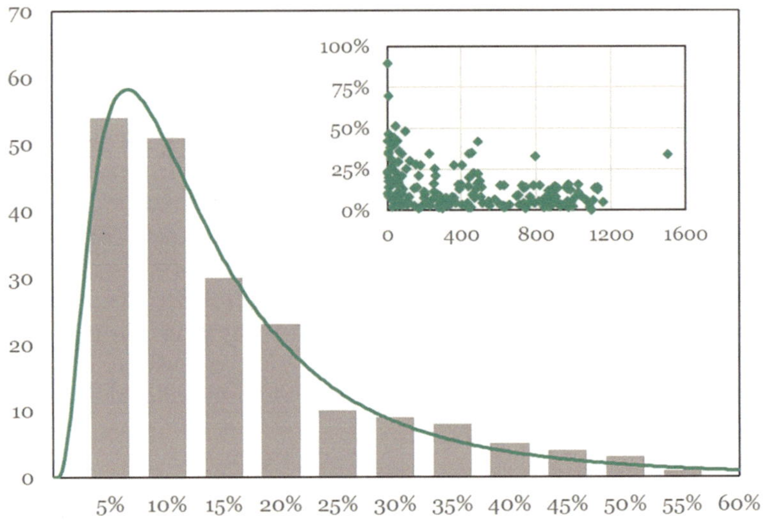

Figure 2.
Histogram and estimated probability distribution for percent difference between individual and averaged strains (inset figure – Percent difference vs. maximum average strain).

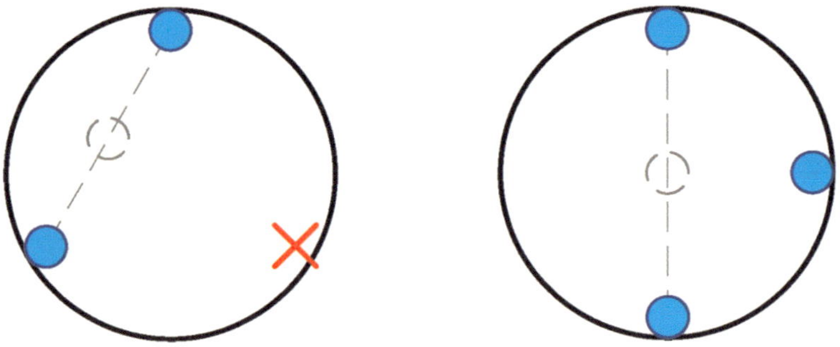

Figure 3.
Strain gage triplet averaging results with defective gage (left), and with 0°, 90°, 180° arrangement (right).

values for this case history). This is because in either arrangement, the average strain at the centroid is successfully computed only if all the gages function, and assuming each individual gage has an equal probability of malfunction, there is a higher cumulative probability of losing one gage out of three installed than one out of two installed.

In this test program the three specified gages were installed at 0°, 90° and 180° around the rebar cage at each level (see **Figure 3**). The gage at the 90° position was logged but the data was not used in the analysis of results unless one of the other gages malfunctioned. This resulted in a slight improvement in the overall test program; five of the seventeen malfunctioning gages were at the 90° position, resulting in no negative effect on the data analysis.

Substantial redundancy is achieved by installing four strain gages per level, if they are viewed as two independent sets of opposed pairs. If all four gages function properly, then the average strain is computed from all four. However, if any one gage malfunctions, it and its opposed twin is discarded and the average is computed from the remaining opposed pair only, which should still yield a good measure of strain at

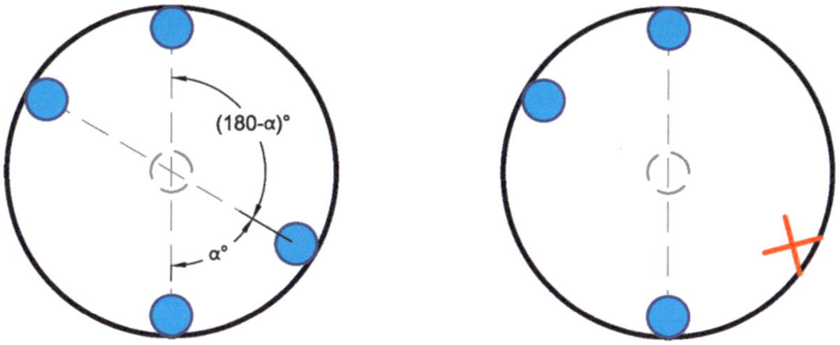

Figure 4.
Averaging results for two opposed pairs of strain gages.

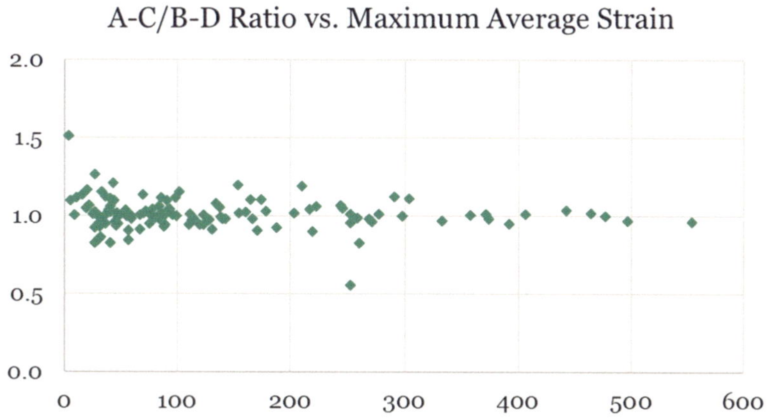

Figure 5.
Scatterplot of ratio of maximum A-C average to B-D average versus overall maximum strain.

the pile centroid. Note that the gages do not have to be spaced at 90° angularly; each pair needs only to be 180° opposed (**Figure 4**).

The probability of success S_{2x2} for this arrangement is computed as one minus the probability of simultaneous failure of both opposed pairs:

$$S_{2\times2} = 1 - (1 - S_2)^2 = 1 - \left(1 - (1 - \lambda)^2\right)^2 \tag{3}$$

For the Florida case history, using the same value λ of 2.6% results in a probability of success of 99.7% (up from 94.9% using two gages in a single opposed pair).

The California/Nevada case history data set consisted of a total of 488 gages in 122 functioning quartets from sixteen drilled shaft tests. By convention, the gages are designated A, B C and D, clockwise in plan around the rebar cage perimeter. The two opposed pairs are then labeled A-C and B-D, respectively. **Figure 5** plots the ratio of the maximum average strain of the A-C pair to the B-D pair versus the average of all four gages.

The average of the ratios is 1.01, indicating that in general the A-C and B-D pairs converge on the same average strain value. However, the standard deviation is 0.10, meaning on average there is a potential for approximately 10% deviation in the measured strain (and thus computed load) using one versus two pairs of gages. Depending on the test objectives, this value may be significant enough to justify specifying four gages per level.

The purpose of embedding strain gages in a test foundation is to determine the load distribution (see below) and from it, the t-z and q-z curves. As such, there are two possible strategies to consider when deciding on the location (depth in the foundation) for each level of strain gages. The first approach will seek to identify the shear capacity of distinct soil layers in the stratigraphy. Based on a nearby (or ideally, centerline) soil boring, gages should be positioned at the interfaces between various soil strata to separately identify the capacity of each. Alternatively, if the test data is to be used as input to a finite-difference computer model such as FB-MultiPier, the gages should be positioned at an even spacing corresponding to the node spacing in the computer model. Consultation with the design engineer during the planning phase of a load test program will help identify test objectives and inform the optimal layout of strain gage levels. As a general guideline, gages should not be located closer than one element diameter to boundaries of the foundation (top, base and/or embedded loading device for bi-directional tests), in order to assure a plane-strain condition.

3. Incremental back-calculation

The function which converts axial strain to stress in a deep foundation is the multiplier consisting of Young's modulus of the foundation material *E* times the cross-sectional area *A*. This function is often called the 'stiffness' of the foundation, although technically this is a misnomer since by definition the units of stiffness are force per length (*AE/L*), whereas the conversion of strain (unitless) to force must also be defined in units of force (*AE*), and is properly called the 'rigidity.' Composite axial rigidity calculations based on empirical relationships such as the ACI 318 formula [3] result in a constant value of *AE*. These types of empirical formulas are based on several assumptions, including average concrete strength *f'c* and knowledge of the cross-sectional area, which may be only nominally correct. In addition, confinement effects and the fact that the stress-strain relationship (modulus) of cementitious materials is not linear are also not considered.

The basic assumption of incremental rigidity back-calculation methods is that the non-linear stress-strain relationship of cementitious materials (drilled shaft concrete and augercast pile grout) can be adequately approximated with a quadratic function [4]. This assumption seems reasonable if a family of stress-strain curves is examined (**Figure 6**), with a parabola (red-dashed line in the figure) overlaid over the *f'c* = 4000 psi curve as an illustrative example. The approximation is quite good from the origin up to the peak stress (yield point), which is all that is required for the analysis of load test results. The value of Young's modulus is the slope of the stress-strain function curve at any given strain. As noted in the case histories in the previous section, it is not uncommon during axial load testing to measure strains on the order of 500 to 1000 $\mu\varepsilon$ or more, especially in slender elements such as ACIP piles. Therefore, the non-linearity of Young's modulus of cementitious materials must be accounted for.

The axial rigidity contribution of steel reinforcement in the composite cross-section is typically relatively small, and the stress-strain curve for steel is assumed linear up to the yield point which means its modulus is relatively constant. Therefore,

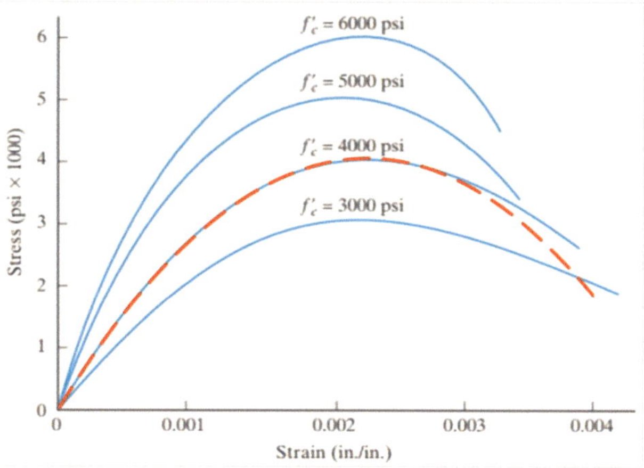

Figure 6.
Concrete stress-strain curves and quadratic approximation (red dashed line).

the non-linear properties of the cementitious material govern the composite cross-section rigidity of the foundation element.

The equation of the parabolic quadratic function illustrated in **Figure 6** is given by:

$$\sigma = a\varepsilon^2 + b\varepsilon + c \tag{4}$$

Note the constant term c is always zero, which assures that the parabola intersects the origin. The cross-sectional area is assumed to remain constant throughout the analysis; the small effect of Poisson's ratio is neglected. Therefore, to convert from stress to force and modulus to rigidity, Eq. (4) is simply scaled by the cross-sectional area A:

$$F = A\sigma = A\left(a\varepsilon^2 + b\varepsilon\right) \tag{5}$$

3.1 Incremental back-calculation

The Secant Modulus (SM) method is the simplest back-calculation technique. A strain gage (or set of strain gages) is installed in the deep foundation element at or above ground level, such that all of the applied force P at the head of the element must be registered by the gage(s). That is to say, no force is shed into the soil via skin friction between the point of load application and the strain measurement. At each incremental step n of the load test, the rigidity is computed as:

$$(AE_n)_{secant} = \frac{P_n}{\varepsilon_n} \tag{6}$$

Note that the method is typically discussed in terms of stress and strain [5, 6]. Starting with test load data, the foundation cross-sectional area is divided out in order to derive a function for the modulus. In order to recover forces, the cross-sectional area must be multiplied back into the analysis later. Herein, this intermediate step is eliminated since the ultimate objective is to convert strain data to force.

This method is called 'secant' because the resulting rigidity function is the slope from any point on the force-strain curve back to the origin. Once testing is complete, each of the rigidity values AE_n are plotted versus ε_n, and linear regression is used to determine the slope and offset (a and b respectively) of the best-fit line through the data. Substituting Eq. (5) into Eq. (6), the result in terms of the quadratic function presented above is:

$$(AE)_{secant} \approx \frac{F}{\varepsilon} = \frac{A(a\varepsilon^2 + b\varepsilon)}{\varepsilon} = A(a\varepsilon + b) \tag{7}$$

To compute the force at any strain, Eq. (7) is rearranged:

$$F_n = A\left(a\varepsilon_n^2 + b\varepsilon_n\right) \tag{8}$$

One drawback of this method is that it cannot be utilized during bi-directional testing. The plane-strain assumption (the strain measured by the gages is an accurate representation of the average strain throughout the cross-section) means that the gages must be positioned at least one element diameter away from the point of loading, in order for local stress variations at the point of loading to even out. In a bi-directional test, significant force may be shed into the soil via skin friction within this span, invalidating the relationship in Eq. (6) because the force at the strain gage location is now an unknown.

Additionally, in cast-in-place foundation elements (with variable cross-section area and curing conditions), or those which have variable reinforcement with depth, the SM method may not yield accurate rigidity estimates for embedded strain gage levels because the ground-level gages may not be representative [7].

3.2 Tangent modulus and incremental rigidity methods

The Tangent Modulus (TM) method was initially derived by Fellenius explicitly for the modulus, with the cross-sectional area considered separately. This is best applicable for foundation elements with assured constant cross-section properties (such as driven piles). The Incremental Rigidity (IR) method discussed in [7, 8] recognized that the rigidity (modulus times area, AE) is a single function which can be identified without explicitly identifying the relative magnitude of either of the two components A and E. In this discussion the analysis focuses on the strain-force relationship (effectively, the Incremental Rigidity method), although from a strictly mathematical derivation standpoint, the TM and IR methods are equivalent.

In the IR method, rigidity is computed as the slope of the force-strain curve at a given strain. This slope is approximated as the change in applied load ΔP divided by change in strain $\Delta\varepsilon$ for successive load increments:

$$(AE_n)_{incremental} \approx \frac{\Delta P}{\Delta \varepsilon} = \frac{P_n - P_{n-1}}{\varepsilon_n - \varepsilon_{n-1}} \tag{9}$$

As with the SM method above, once testing is complete the rigidity value at each increment is plotted against its corresponding strain and a best-fit line plotted through the data. However, the incremental method requires that the side shear section between the point of load application and the strain gage elevation has reached or at least approached its ultimate capacity. Because of shear resistance, the force increase

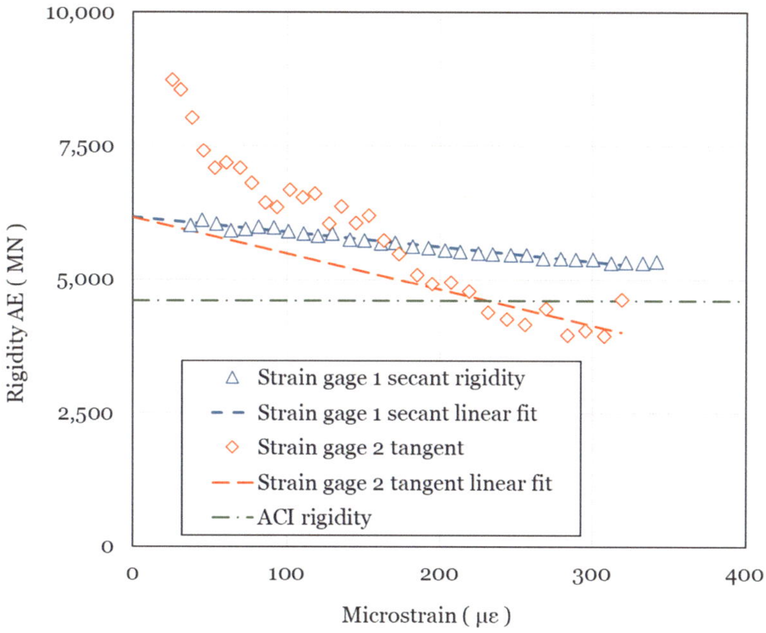

Figure 7.
Sample rigidity analysis from strain gage data.

at the strain gage will be less than the applied load increase during the initial part of the test. Therefore, the resulting incremental rigidity values will be excessive. It is only after the side shear section between the point of load application and the strain gage reaches its ultimate shear capacity that subsequent applied load increments result in strain increments which give a true indication of the rigidity. This behavior becomes apparent on a plot of the analysis where the rigidity decreases from very high values at a small ε to a linear curve at high ε (see **Figure 7**, above). A linear regression through this ultimate portion of the incremental rigidity curve will yield slope and offset values g and h:

$$(AE)_{incremental} = A(g\varepsilon + h) \tag{10}$$

The incremental rigidity, by definition is also the slope (first derivative) of the force-strain function (Eq. (5)):

$$(AE)_{incremental} = {}^{dF}\!/_{d\varepsilon} = A(2a\varepsilon + b) \tag{11}$$

Comparing Eqs. (7), (8) and (11) it becomes apparent that the incremental rigidity and secant rigidity analyses for the same load test will result in a different force-strain relationship, by a factor of 2 in the first term (slope) and that the second term constants (b and h respectively) are equivalent. This is illustrated in **Figure 7** with a sample data set from a series of top-down tests (the 'Texas' case history). Strain gage 1 is located just below the point of load application, and is analyzed using the secant rigidity method. Strain gage 2 is located some distance down within the shear embedment zone, and is analyzed using the incremental rigidity method. As expected from theory, the two linear regressions converge at the vertical axis (the zero-strain condition) but have significantly different slopes. For comparison purposes, the ACI

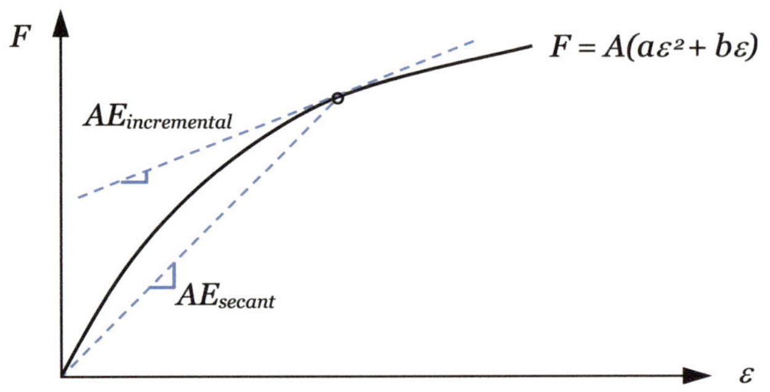

Figure 8.
Non-linear force-strain curve with incremental (tangent) and secant moduli.

rigidity (computed using the empirical relationship to the square root of concrete strength *f'c*) is also plotted as a horizontal line, showing that it does not produce a good result for this particular foundation element.

Because it is the slope from any point on the force-strain curve back to the origin, the secant rigidity can be multiplied by any measured strain to directly compute force. However, the incremental rigidity cannot be simply multiplied because it is by definition tangent to the force-strain curve at all points and does not intercept the origin. **Figure 8** illustrates this point graphically.

By simply multiplying the curve-fit incremental rigidity slope by 0.5, the equivalent secant rigidity function is recovered and Eq. (12) used to compute the force directly for each measured strain.

$$F_n = A\left(0.5\left(g\varepsilon_n^2\right) + h\varepsilon_n\right) = A\left(a\varepsilon_n^2 + b\varepsilon_n\right) \tag{12}$$

Alternatively, or when dealing with highly non-linear rigidity relationships, the value of F at any loading point n may be approximated by a recursive summation formula [9]:

$$F_n = F_{n-1} + (AE_n)_{incremental}(\varepsilon_n - \varepsilon_{n-1}) \tag{13}$$

where F_{n-1} and ε_{n-1} are the force and strain of the previous loading data point, respectively. This step-wise approximation will roughly follow the curved load-strain path.

This approach will give approximately correct results even if the rigidity function is highly non-linear, such as in the case of a tensile load test once the cementitious material begins to crack due to tensile strain, or in a compressive load test with pre-existing tension cracks in the cementitious material which are closed up by the compressive axial stress [9]. As noted above, foundation element axial rigidity AE is composed of two contributors, steel and cementitious material (A_sE_s and A_cE_c, respectively). For a cementitious material which is fractured (due to shrinkage during curing or applied tensile stress), the nominal area A_c is replaced with an effective area A'_c.

Figure 9 illustrates two idealized functions of nonlinear axial rigidity due to cementitious material fracturing in response to tensile stress (bold line segments). The full composite rigidity consists of $A_sE_s + A_cE_c$. The angular pathways to/from the reinforcing steel rigidity (A_sE_s only) indicate idealized changes in rigidity due to fracturing with increasing strain. With increased compressive strain pre-existing

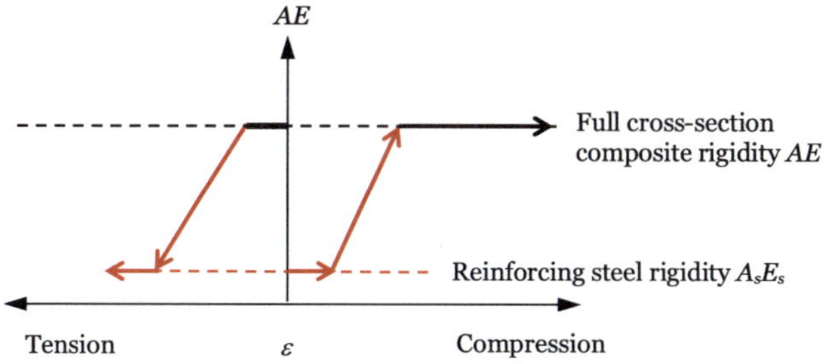

Figure 9.
Possible non-linear rigidity function paths.

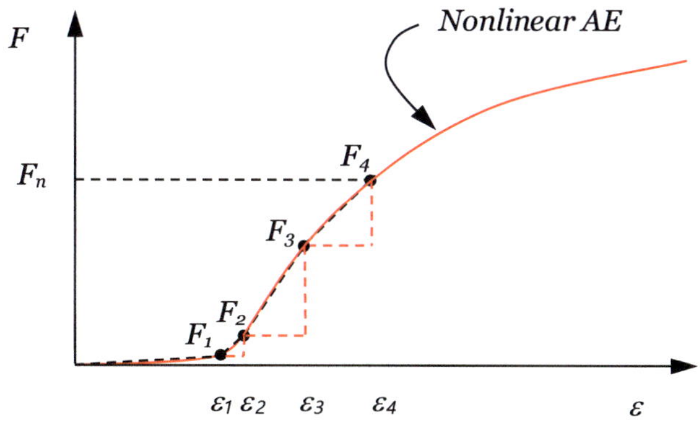

Figure 10.
Incremental load calculation.

fractures progressively close up, A'_c increases from zero to A_c and the rigidity increases until the full composite rigidity is reached. Conversely, with increased tensile strain, the rigidity decreases from the full composite value down to the reinforcing steel rigidity only, as the cementitious material progressively fractures until only the reinforcing steel remains to transmit stress.

If Eq. (13) is employed, using each load test increment as a discrete step the non-linear load-strain curve can be approximated by a series of small incremental increases in load, each of which is linear with its corresponding increase in strain, as illustrated in **Figure 10**.

Note that all the rigidity back-calculation methods depend on obtaining high-quality strain gage data from relatively small, equal load increments to clearly define trends. Results obtained at one strain gage level may not apply at other levels, due to several factors including possible changes cross-sectional area, reinforcement details, confinement (within rock socket as opposed to overburden) and differing concrete curing conditions (hydrostatic pressure, water table elevation, environmental temperatures, etc.) among others.

Once the load at each strain gage level has been computed using the methods discussed above at every load increment, a family of load distribution curves can be generated (see **Figure 11**).

The difference between adjacent levels (a 'zone' of the foundation element), divided by the perimeter shear area of the zone, gives the unit shear, the 't' component of the desired t-z curve. A level of strain gages placed near the base of the foundation also allows for estimation of the bearing resistance q.

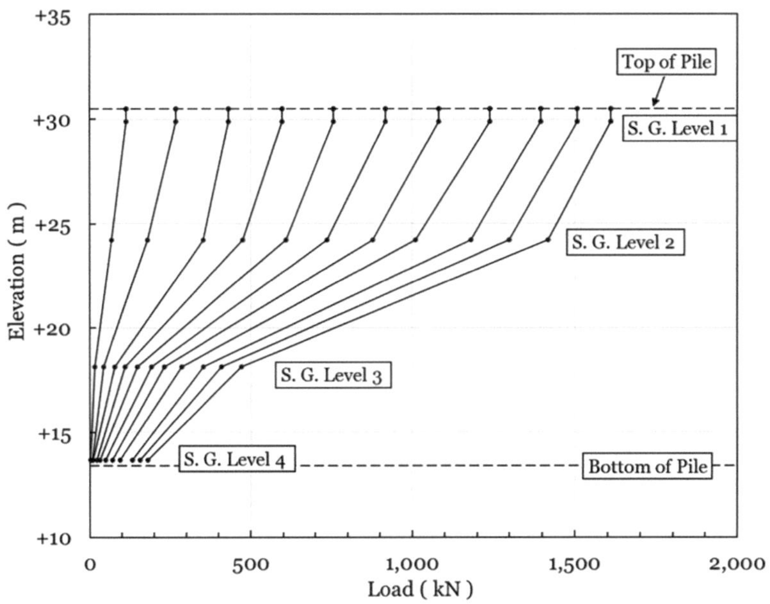

Figure 11.
Sample load distribution (Texas case history).

Note that the analysis results presented herein are based on re-zeroing all strain gages prior to the start of loading, and account for the resistance of the as-built isolated test element to a relatively short-duration externally-applied load only. They do not account for any residual load present in the element at the start of testing, down-drag, long-term setup, creep or group effects.

4. Compression and displacement

To a first-order approximation, the displacement of each zone of an axially-loaded deep foundation is the measurement D given by dial gages or displacement transducers at the head, or point of load application in bi-directional load tests. However, as discussed above the foundation element has a rigidity, which means it will compress or elongate elastically under applied stress. The degree of this compression or elongation can be estimated using the collected strain gage data.

For each zone, the zone strain is computed as the average of the measured strains at the top and bottom of the zone. Change in length (compression or elongation) δ is then computed as the average zone strain times the zone length L.

$$\varepsilon_{zone} = \frac{\varepsilon_{top} + \varepsilon_{bottom}}{2}, \delta_{zone} = \varepsilon_{zone} \cdot L_{zone} \tag{14}$$

Zones which do not have strain gage levels both at the top and bottom, but rather are located next to boundary changes (the zone(s) adjacent to the load-application device and/or the top and bottom of the foundation element) must be evaluated differently. Depending on the situation, the one available strain gage level may be assumed to be representative of strain throughout the zone, strain data from two or

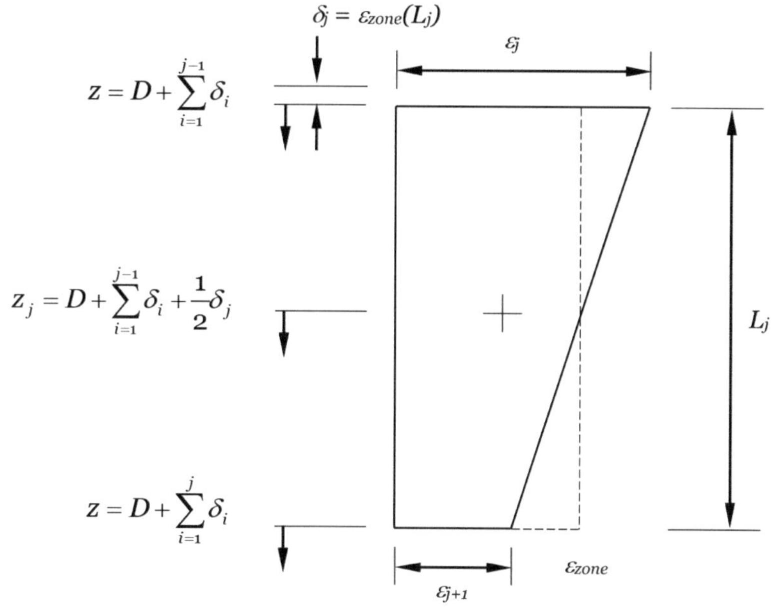

Figure 12.
Calculation of average strain, compression and displacement at top, mid-point and bottom of a zone.

more levels may be extrapolated, or strain may be estimated by correlation to extensometer telltale rod data.

The total displacement of each zone z_j is then computed at the midpoint of the zone. The calculation must include the displacement at the point of load application D as well as the change in length of all zones between the point of load application and the current zone:

$$z_j = D + \sum_{i=1}^{j-1} \delta_i + \frac{1}{2}\delta_j \tag{15}$$

This calculation will then result in the 'z' component of the t-z curve for every shear zone. **Figure 12** is a schematic of the zone displacement calculation.

In certain circumstances, the elastic compression of the test foundation may be a minor contributor to the total computed displacement. However, in situations with very stiff soils or rock, and/or with slender elements with a relatively small rigidity and large strains, the elastic deformation can be a significant if not major portion of the total displacement of each shear zone. While the interpretation of data described in this section is simpler than in the previous sections, it is no less critical to constructing the t-z and q-z curves from strain data correctly.

5. Conclusions

Strain gages are critically important instruments for monitoring the performance of a deep foundation element undergoing a load test. Strain gage data must be properly analyzed in order to gain insight into the true soil-structure interaction. In order to verify design assumptions and particularly to optimize computer models, the unit capacity curves (so-called t-z and q-z curves) must be obtained from the load test. Properly analyzed strain gage data contributes to both components of these curves. However, strain gages are not load cells – the conversion of strain to load is not as straightforward as linear elastic theory may lead one to believe. Careful attention must be given to the selection, placement, monitoring and interpretation of strain gages in deep foundation load testing. The techniques described herein have been developed and successfully deployed by the author and others for over two decades to utilize strain gages in deep foundation load testing.

Author details

Jon Sinnreich
Load Test Consulting Ltd., Gainesville, FL, USA

*Address all correspondence to: jon@ltc-us.com

References

[1] Meyer PL, Holmquist DV, Matlock H. Computer predictions for axially-loaded piles with non-linear supports. In: 7th Offshore Technology Conference. Houston, TX: American Petroleum Institute; 1975. pp. 375-387

[2] Sinnreich J. Optimizing the arrangement of strain gauges in pile load testing. ASTM GTJ. 2020;**44**(5): 1552-1558. DOI: 10.1520/GTJ20200033

[3] ACI. Building Code Requirements for Structural Concrete (ACI 318–14) and Commentary. American Concrete Institute. Michigan, USA: ACI World Headquarters: Farmington Hills. 2014

[4] Sinnreich J. A discussion of Back-calculated rigidity methods. In: DFI SuperPile '22. St. Louis, MO: Deep Foundation Institute; 2022

[5] Fellenius BH. Tangent modulus of piles determined from strain data. In: ASCE Geotechnical Division 1989 Foundation Congress. Vol. 1. Evanston IL: American Society of Civil Engineering; 1989. pp. 500-510

[6] Fellenius BH. From strain measurements to load in an instrumented pile. Geotechnical News. 2001;**19**(1):35-38

[7] Komurka VE, Robertson S. Results and lessons learned from converting strain to internal force in instrumented static loading tests using the incremental rigidity method. In: ASCE Geo-Congress 2020. Minneapolis MN: American Society of Civil Engineering; 2020. pp. 135-152

[8] Komurka VE, Moghaddam RB. The incremental rigidity method – More-direct conversion of strain to internal force in an instrumented static loading test. In: ASCE Geo-Congress 2020. Minneapolis MN: American Society of Civil Engineering; 2020. pp. 124-134

[9] Sinnreich J. Strain gage analysis for nonlinear pile stiffness. ASTM GTJ. 2012;**35**(2):367-374. DOI: 10.1520/GTJ103412

Chapter 3

Experimental Investigation of Glass Fiber Reinforced Clayey Soil for Its Possible Application as Pavement Subgrade Material

Suchit Kumar Patel

Abstract

A clayey soil reinforced with commercially obtainable20 mm glass fiber of varying fiber content (f_c = 0.25 to 1% by soil dry weight) was investigated in lab for its possible application as road pavement material. Standard proctor compaction, unconfined compression strength (UCS), California Bearing Ratio (CBR) and undrained triaxial compression tests were conduction on compacted soil-fiber specimens as per ASTM standard. From the fiber mixing process it has been observed that fiber can be uniformly mixed into clayey soil only up to some optimum fiber content. Laboratory test results predicted that UCS, CBR and shear strength value of clayey soil enhanced significantly with fiber content up to some optimum value of 0.75% fiber content. The UCS increases maximum up to two fold, CBR by 2.8 times and shear strength by around 1.75 times than that of clayey soil alone. The inclusion of glass fibers enhances the ductility of clayey soil and modifies its failure pattern from brittle to ductile. It has been found that the glass fiber reinforced clayey soil can be used significantly as a subgrade material for low volume flexible road pavement.

Keywords: glass fiber, UCS test, CBR test, shear strength, subgrade material

1. Introduction

The increasing need of construction areas for infrastructure facilities like roads, rails, etc. that spread over large spaces confines the choice of neglecting those locations where poor soil is available. This open the scope of strengthening the available soil by using some ground improvement method. Ground improvement methods address many ground conditions problem and help in modifying the engineering aspects of available soils as per the requirements. These techniques also help in obtaining economical and environmental friendly solutions to mitigate the issues related to soil for construction purpose. Some basic ground improvement techniques including densification, dewatering and use of admixtures and reinforcement are being adopted from ancient times.

1.1 Ground improvement methods

Several ground improvement techniques are currently in use to improve the engineering properties of soils. These modification techniques have been divided into several categories [1].

a. Mechanical modification: This includes physical modification of soil and can be carried out by means of controlled densification either by placement and compaction of soil or in-situ methods of soil improvement for deeper application. This includes static and dynamic compaction, and vibro-compaction. This method is most suitable for granular soils.

b. Hydraulic modification: This involves the modification of flow, seepage and drainage characteristics of soil. This is done by lowering water table, decreasing or increasing soil permeability, consolidation and preconsolidation by using vertical drains to minimize settlement and compressibility and increasing overall strength.

c. Physical and chemical modification: This deals with the stabilization of soil by physiochemical changes of the soil structure. This includes physical mixing of some chemical or additive material like cement, lime, industrial wastes (fly ash, ground granulated blast furnace slag etc.), injection of grouting materials, bioremediation of soil, thermal treatment.

d. Modification by inclusion, confinement and reinforcement: This includes application of some other manufactured materials within the soil mass. This involves use of reinforcement as tension resisting element in different forms known as soil-reinforcement. This also includes soil nailing, soil anchor and inclusion of stone column.

The above mentioned soil modification techniques are not limited to any particular type of soil. It can be adopted for any soil depending on their suitability and ease of field applicability, economic constraints along with the availability of resources for their implementation on any particular site.

1.2 Soil reinforcement

Among several ground improvement methods, soil reinforcement is an effective and dependable method for upgrading the strength and stability of various civil engineering construction practice including pavement, embankment, retaining structures, foundations and slopes. Reinforced soil is a composite mass in which tension resisting elements in different forms (geosynthetics, fibers etc.) are embedded to increase the strength, stiffness, compressibility and permeability of soils. After the earliest reinforcement in the form of galvanized steel strips of high tensile modulus, use of synthetic materials named as geosynthetics in different forms (geogrid, geotextile, geocomposite etc.), and of natural products (bamboo, jute, and coir) are being adopted in the form of sheets or meshes. In most applications, the conventional method of soil reinforcement is in a continuous planer form introduced within the soil mass in a definite pattern, resulting in the systematically reinforced soil [2].

The one-dimensional orientation of reinforcement is installed sequentially in alternating layers as per the design requirements of the structure.

1.3 Fiber-reinforced soil

Fiber-reinforced soil has gained popularity in around last 35–40 years [2] where flexible, discrete fibers are being mixed within soil mass. Fibers act like tension resisting element which cause significant amendment in the various engineering aspects of soil including strength, stiffness, compressibility, permeability. Unlike conventional soil reinforcement methods, fiber-reinforced soil maintain strength uniformity within the soil mass by evading the generation of any weak plane during field placement. Fibers are available in abundance in natural and waste form, and also manufactured in desired properties known as synthetic fibers. Utilization of waste fibers (tyre derived fibers, plastic waste fibers etc.) for civil engineering work can help in solving disposal problems which will be cost effective and also help in enriching the environment.

The method of fiber reinforcement in soil is being used from ancient times where natural fibers in the form of straw were mixed in the soil brick to provide integrity by arresting the crack development [3]. The curiosity of fiber-reinforced soil in last century started by Waldron [4] when he investigated the effect of roots of plant and tree on the earth slope stability. With increasing attention, fiber reinforced soil is increasingly providing an option of its use behind retaining structure as backfill material [2], construction of embankments [5–7], slopes stabilization [8], earth retaining constructions [9] and clay liners [10].

The use of fibers in natural and synthetic form like coir, jute, wool, steel, nylon, polyester, polypropylene, and fiber glass as tension elements for various soil have been reported by other investigators by means of unconfined compression, CBR, direct shear and triaxial compression tests in the last 35–40 years [2]. However, the preliminary works was largely on fiber-reinforced sand where the influence of the key aspects such as fiber concentration, fiber aspect ratio, soil compaction level and testing environments on the overall performance of fiber-reinforced sand was studies [11–15].

The effects of fiber inclusion on clayey soils have been explored by direct shear tests [16–18], triaxial compression tests [19–22], unconfined compression tests [23–29], tensile strength tests [30], fiber pullout tests [31] and CBR tests [32–34]. The common findings of the past investigations on fiber-reinforced soil are that the fiber inclusion increases the stress–strain responses, UCS, soil ductility and CBR, and modify the post-peak strength reduction of soil. The inducement of shear strength happens up to some controlling fiber concentration and fiber length.

The fiber benefits depend on the bond strength and surficial interaction between soil and fiber [17]. The soil particle size also influences the shear strength of fiber reinforced soil [35]. Fiber reinforcement had also effectively reduced the amount and degree of desiccation and tension cracks development, suppressed the swelling potential, and increased the permeability of clay soils [36–38]. It has been noted that the compressive strength of fiber-reinforced soil is highly controlled by the size of specimen [39] and compaction state [28, 29].

In this present study, an attempt has been made to investigate the effect of glass fiber inclusion on the strength aspect of a clayey soil for its possible suitability for road pavement construction. The investigation has been carried out by conducting compaction, UCS, CBR and triaxial compression test by varying fiber content.

2. Materials and methodology

2.1 Soil

Locally available clayey soil was found form the nearby hill slope in the outskirt of Guwahati city of Assam state in India. The particle size distribution curve of the tested soil is presented in **Figure 1**. The soil confined 25%, 54% and 21%, sand, silt and clay size particles, respectively. The soil had 46% liquid limit, 25% plastic limit value. As per Unified Soil Classification System (USCS) according to ASTM D2487 [40], the soil was classified as low plastic clay (CL). The coefficient of uniformity and the coefficient of curvature based on the gradation curve were 12.5 and 3.125, respectively. The optimum moisture content (OMC) and maximum dry unit weight (MDU) values of the soil were 19.4% and 16.8 kN/m^3, respectively as per ASTM D698 [41].

2.2 Reinforcement

Glass fiber of 20 mm length nd 0.15 mm average diameter was used as reinforcement (**Figure 2**). The glass fiber has specific gravity and water absorption capacity as 2.57 and zero respectively. The modulus of elasticity, tensile strength and elongation at break of the glass fiber were 112.3 GN/m^3, 1.53 GN/m^2, and 1.8%, respectively. As glass fiber has higher stiffness, strength, high ratio of surface area to weight, dimensional stability [42], and is ready available and non-biodegradable [43], it can be more valuable for long-term soil remediation. Glass fiber has also been found to retain its elastic modulus and tensile strength at 70–75% of that of raw fibers even under 450°C temperature [44] and thus will be suitable for the country where environmental temperature becomes high in the range of 50° in summer.

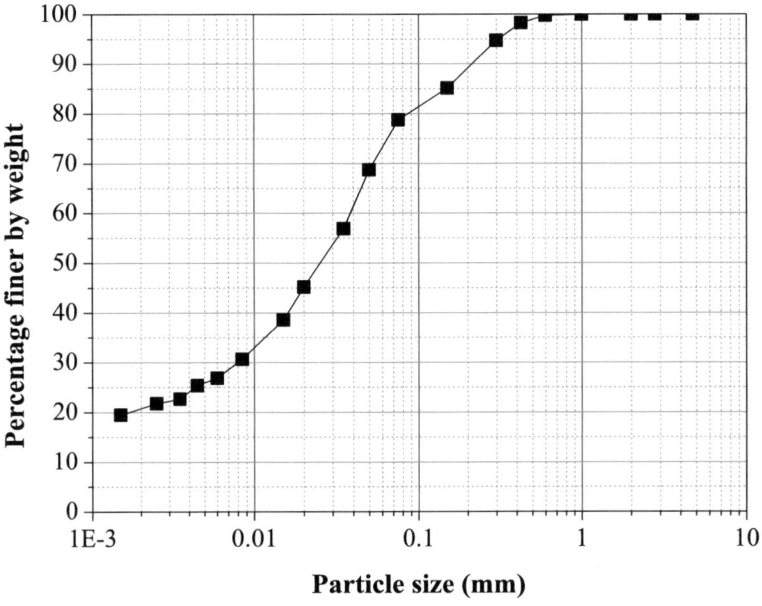

Figure 1.
Particle size distribution curve of clayey soil.

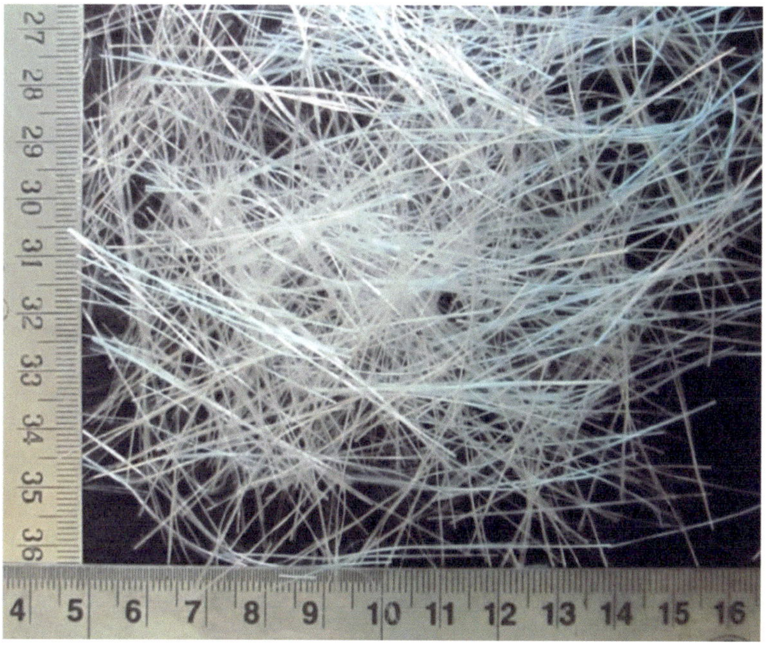

Figure 2.
Commercially available glass fiber used in this study.

2.3 Specimen preparation

Designated weight of dry soil, fiber and water was taken and mixed in a steel tray. At first, the dry soil was mixed only with water, and then fiber was added with moist soil in small increments manually taking proper care. Thereafter, the soil-fiber homogeneous mix was shifted to a poly bags and reserved in a desiccators for 24 hrs to confirm its moisture steadiness. Afterward, the soil-fiber mixture was compacted in a cylindrical mold of 38 mm inner diameter having detachable collars at both ends for UCS and triaxial test sample. The whole amount of moist soil-fiber mix was shifted into the mold from either end, after fixing the collar at the other end. Subsequently, compaction was done from both ends by giving simultaneous equal rotation to the collars till the specimen length of 76 mm was attained. For CBR test, the specimen was compacted in CBR mold using standard proctor compaction energy as per ASTM D 698 [41].

It was decided not to go for fiber content above 1% as with 1% fiber content, homogeneous mixing of fibers was difficult due to formation of soil-fiber lumps. 20 mm fiber of different fiber doses (f_c = 0.25, 0.5, 0.75, and 1% by parent soil dry weight) were selected to mold the soil-fiber samples.

2.4 Testing programme

The standard compaction tests were performed for unreinforced and glass fiber-reinforced soil according to ASTM D698 [41] to obtain the OMC and MDU value of various mixes. Unconfined compressive Strength (UCS) test were performed as per ASTM D 2166/D 2166 M [45] with 1.25 mm/min axial strain rate for all specimens. Consolidated undrained (CU) triaxial tests were performed according to ASTM

D4767 [46] with an axial strain rate of 0.12 mm/min for different soil fiber mixes under varying confining pressure ranging from 100 to 400 kPa. Load, axial deformation and pore pressure during triaxial test were electronically measured and recorded by load cell of a capacity of 10 kN with a sensibility of 0.01kN, LVDT of capacity ±20 mm with a sensibility of 0.01 mm and pore pressure transducers, respectively. The CBR tests were performed as per ASTM D 1883 [47] under both unsoaked and soaked conditions for all soil-fiber mixes.

3. Results and discussions

3.1 Mixing efficiency of fiber

The effectiveness of fiber within soil depends on its mixing efficiency. To investigate the distribution of fibers along the height of reinforced specimen, several fiber-reinforced specimens were disintegrated along its height. Three individual specimens were prepared for each fiber length and fiber content, and each specimen was cut into three equal pieces along the specimen height and the weight of fiber in each piece was calculated. At the time of specimen cutting along diameter, it was noticed that most of the fibers within specimen were aligned in the near horizontal direction perpendicular to the specimen height. Further, the fibers were noted to be uniformly distributed in the cutting plane of each specimen.

For segregating the fibers from the soil-fiber mix, each piece was crushed separately and the crushed soil-fiber was washed through a net of sieves of size 2 mm, 0.425 mm and 0.075 mm. All the soil particles were completely washed away from the 2 mm sieve to the 0.425 and 0.075 mm sieves, whereas most of the fibers were retained on 2 mm sieve. Further, the retained materials on 0.425 and 0.075 mm sieves were transferred to a bucket containing water. Then the water was stirred which settled the soil particles and fibers were accumulated on the water surface. In this way, the fibers were completely separated from the soil-fiber mix from each individual piece. The collected fibers of individual piece were oven dried and weighted. The percentage of fiber in each piece of individual specimen was then evaluated based on total weight of fiber mixed in that specimen.

Typical values of measured fiber content in three different parts of specimens of UCS test for different soil-fiber mixes are given in **Table 1** along with their standard deviation. The percentage of distributed fibers within three different parts of any reinforced specimen is relatively close. Therefore, it can be inferred that the mixing

L, mm	f_c, %	1/3 Top		1/3 Middle		1/3 Bottom	
		Fiber distribution, %	Standard deviation, ± %	Fiber distribution, %	Standard deviation, ± %	Fiber distribution, %	Standard deviation, ± %
20	0.25	33.37	0.80	34.60	1.61	32.00	0.16
	0.5	34.07	1.03	32.07	2.29	33.90	0.51
	0.75	32.77	1.60	34.33	2.58	33.80	2.16
	1	33.43	2.35	33.37	3.56	33.20	4.01

Table 1.
Distribution of fibers in different part of reinforced specimen.

efficiency of fibers is uniform along the height of the specimen to some extent, and fibers can be considered to be distributed homogeneously in the specimen. The standard deviation of fiber distribution is varying between ±0.16 to ±4.01%, and the values are found to be higher at higher fiber content indicating that the fiber mixing efficiency decreases at higher content.

However, during field application, ensuring the uniformity of fiber in the large soil-fiber mass will be very challenging, especially for higher fiber dose. Therefore, for maintaining the uniformity of fibers within soil mass for large scale applications, it is important to use better mixing technique.

3.2 Compaction test results

The compaction curves of all specimens, with different combinations of fiber content are depicted in **Figure 3**, and their respective OMC and MDU values are shown in **Figure 4**. The OMC and MDU of the unreinforced soil are found as 19.4% and 16.80 kN/m^3, respectively. As the fiber content increases, there is a minor enhancement in OMC from 19.4% to 19.7% and a small decrease in MDU from 16.80 to 16.57 kN/m^3. As the OMC and MDU variation is marginal, for specimen preparation of either unreinforced or fiber-reinforced soil, the specimens were compacted at the OMC and MDU value of unreinforced soil.

3.3 UCS test results

Figure 5 presents unconfined compression test curve showing the effect of fiber content for all reinforced specimens. As fibers are added to the soil, the stress–strain behavior has modified appreciably in terms of both peak stress and strain improvement. This is followed by decrease of post-peak stress loss, showing stimulation of plastic nature to the soil and the brittleness nature transforms gradually to ductile.

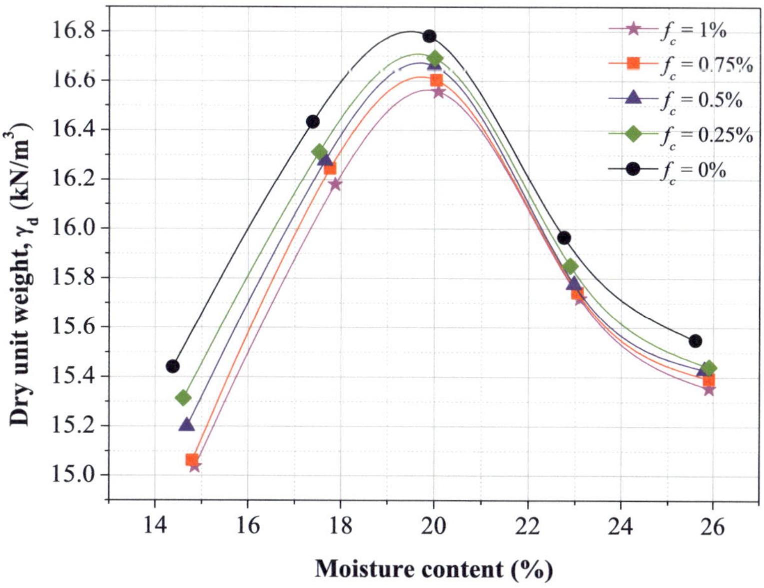

Figure 3.
Effect of fiber content on compaction curve.

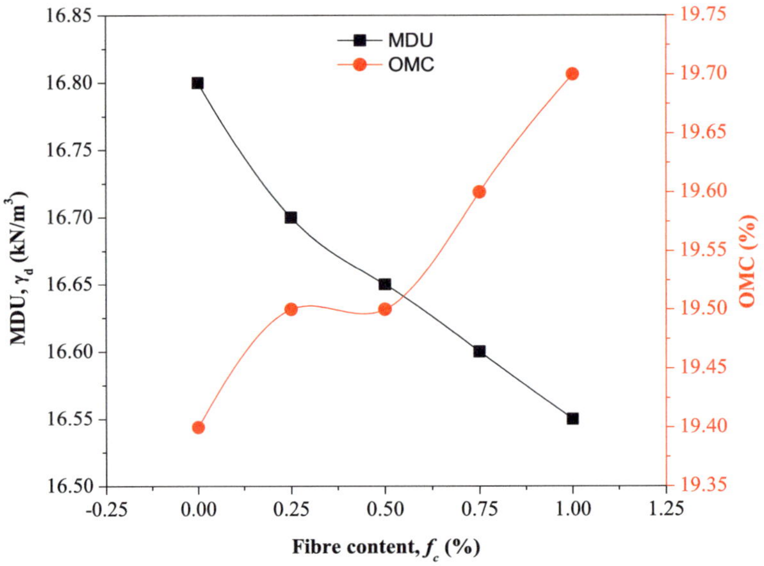

Figure 4.
Variation of compaction parameters (OMC and MDU) with fiber content.

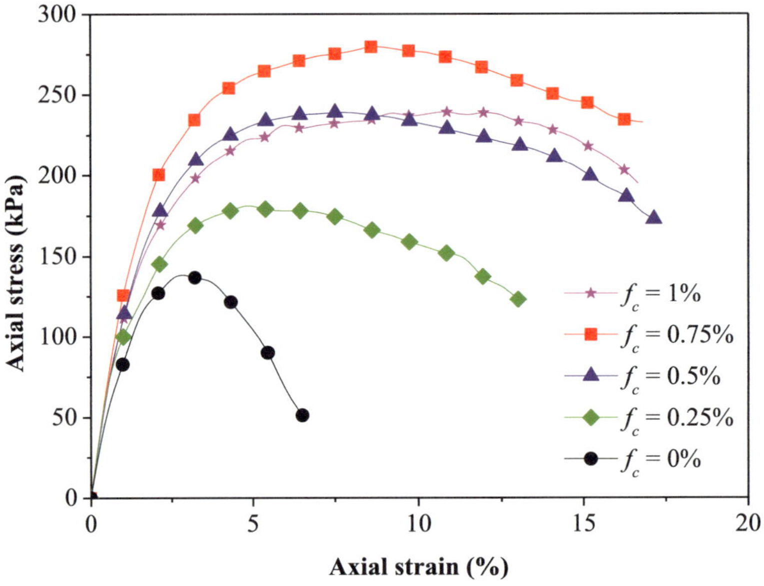

Figure 5.
Effect of fiber content on stress–strain response.

The maximum stress is found for the specimen with 0.75% fibers, and addition of additional fiber of 1% results in strength reduction. This shows that there is an optimal fiber content where advantage of reinforcement is the maximum. As the fiber content increases further to 1%, the number of fibers in soil increases such that the availability of soil matrix quantity for holding the fibers may not be that adequate to develop optimum bond among all soil-fiber interfaces. Consequently, the tensile strength of all fibers is not mobilized completely causing in peak strength drop at

1% fiber. However, the UCS of specimen reinforced with 1% fibers is higher than that of with 0.5% fibers. Fiber reinforcement advantage is mainly subjective to the bond strength and friction between soil particles and fibers [21]. It was also noted that at the time of soil-fiber mixing with 1% fibers, uniform mixing of fibers was difficult and development of fiber lumps started to become visible which hindered the specimen uniformity.

The peak UCS and corresponding axial strain of all tested samples are represented in **Figure 6**. It has been found that with increasing fiber content the peak axial strain is increasing continuously indicating the more ductility in the soil specimen with added glass fibers. The peak axial strain of unreinforced soil was 2.65% which has increases maximum to 10.85% at 1% fiber content indicating around four time increment of peak axial strain. The UCS value is noted to be 137 kPa for unreinforced soil which improved to 181 kPa, 238 kPa, 279 kPa and 239 kPa for 0.25%, 0.5%, 0.75% and 1% fiber content, respectively showing around a maximum two fold increment of UCS value with 0.75% fiber content.

Figure 7 depicts the failure patterns of unreinforced and reinforced specimens. The unreinforced soil specimen (**Figure 7a**), showing a single shear plane across the specimen indicating its brittle behavior. This brittleness of unreinforced soil can also be observed from the stress–strain curve (**Figure 5**), where a sudden drop in stress is noted after peak. For specimen reinforced with 0.25% and 0.5% fiber, some dissimilar multi-shear planes in some portion of the sample are noted to develop (**Figure 7b** and **c**). Whereas, with 0.75% and 1% higher fiber dose, the specimens undergone largely bulging with the development of minor fissures around the sample (**Figure 7d** and **e**). The bridging effect of the fibers restricted the progress of shear planes or fissures, causing reallocation of stresses within the reinforced sample. It has also been noted in stress–strain response that the specimen fails at gradually higher axial strain with high fiber content (**Figure 5**), reflecting the inducement of ductility.

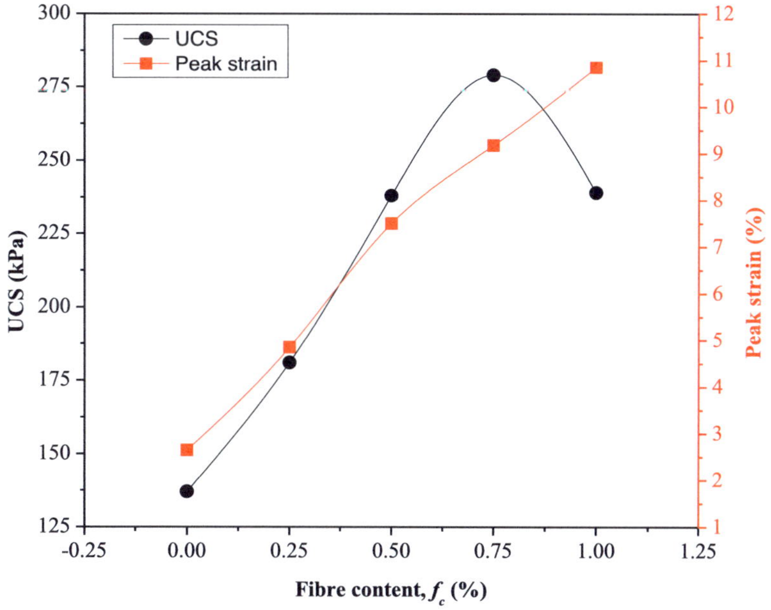

Figure 6.
Effect of fiber content on UCS and peak strain.

Figure 7.
Effect of fiber inclusion on specimen failure mode: (a) f_c = 0%; (b) f_c = 0.25%; (c) f_c = 0.5%; (d) f_c = 0.75%; (e) f_c = 1%.

3.4 CBR test results

The load-penetration responses of the CBR tests on unreinforced and reinforced soil samples with varying fiber content are presented in **Figure 8** for unsoaked condition. The load carrying capability of the samples increases with fiber content up to 0.75%, signifying that fibers can improve the load-penetration behavior. The bearing capacity of the specimens improves continuously with penetration depth up to 15 mm for all fiber contents, representing clearly that the specimen peak strength has not been attained even at 15 mm deformation, and that the fibers have not been pullout or rupture and are still in tension. At higher penetration, the curve slope decreases signifying that the rate of bearing capacity enhancement is diminishing.

The fiber indentations due to the soil particles permit to develop adhesion within soil and fiber [48], ensuring enhanced load carrying capacity of the reinforced soil. Tang et al. [21] told that randomly distributed fibers perform as a three-dimensional

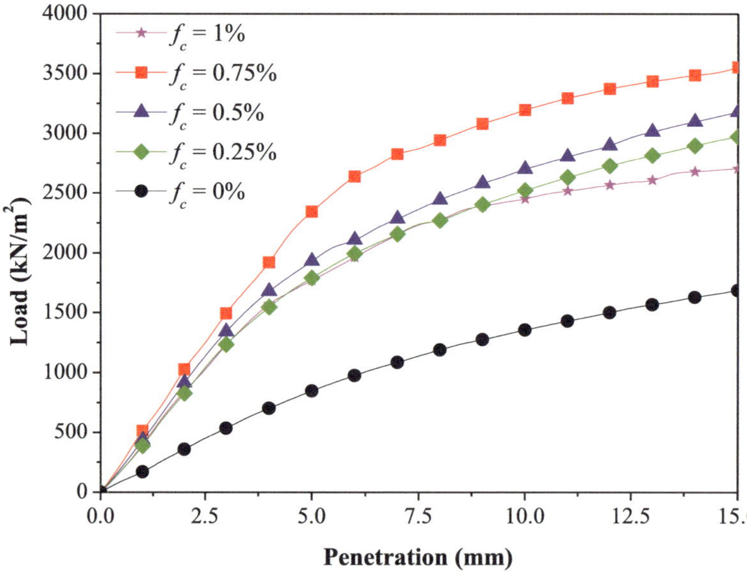

Figure 8.
Effect of fiber content on load-penetration response under unsoaked condition.

arrangement which interlocks soil grains, and restricts the movement of soil, improving the stretching resistance between soil and fibers, ensuing strength inducement. Also, the tensile restraint in the fibers imparts supplementary soil confinement [49] and results in enhancement of specimen strength.

The CBR values under both soaked and unsosked condition are shown in **Figure 9**. Maximum enrichment in CBR for soaking condition is with 0.75% fiber. The maximum enhancement of CBR is from 6.45% to 18.94% under unsoaked condition and 2.89% to 8.23% under soaked condition with 0.75% fiber. For use in field, the determination of optimal soil-fiber mixture is important. For 4 days soaked condition, the CBR of the parent soil is 2.89%, and the maximum CBR of 8.23% is obtained with 0.75% fibers. Therefore, according to IRC: SP: 72 [50], the unreinforced soil is of very poor quality subgrade material (soaked CBR less than 3%), which can be upgraded to good quality subgrade material (soaked CBR between 7% and 9%). However, according to IRC: 37 [51], a minimum soaked CBR value of 6% is essential for subgrade layer of low-volume flexible pavements. Thus, the clayey soil mixed with 0.5, 0.75% and 1% glass fibers having CBR values of 6.89%, 8.23% and 7.62%, respectively can be used in subgrade layer of low-volume flexible pavements.

3.5 Triaxial test results

The effect of fiber content on stress–strain and pore water pressure-strain behavior for all specimens sheared under 100 kPa confinement, are shown in **Figure 10** and **Figure 11,** respectively. The deviator stress-axial strain response was found to enhance continuously with fiber content only up to 0.75% and then remain close to 0.75% fibers with 1% fiber. No peak appears till 20% strain for any specimen tested (**Figure 10**). Similar stress–strain response on fiber reinforced soil where no clear peak was observed, even at an axial strain of 20% was noted by Andersland and Khattak [52], Ranjan et al. [35] and Estabragh et al. [22].

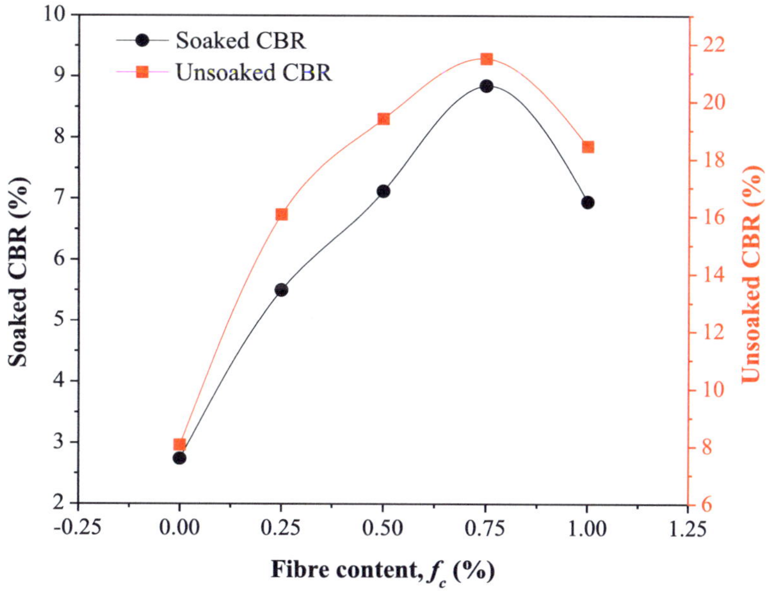

Figure 9.
Effect of fiber inclusion on CBR value under both soaked and unsoaked conditions.

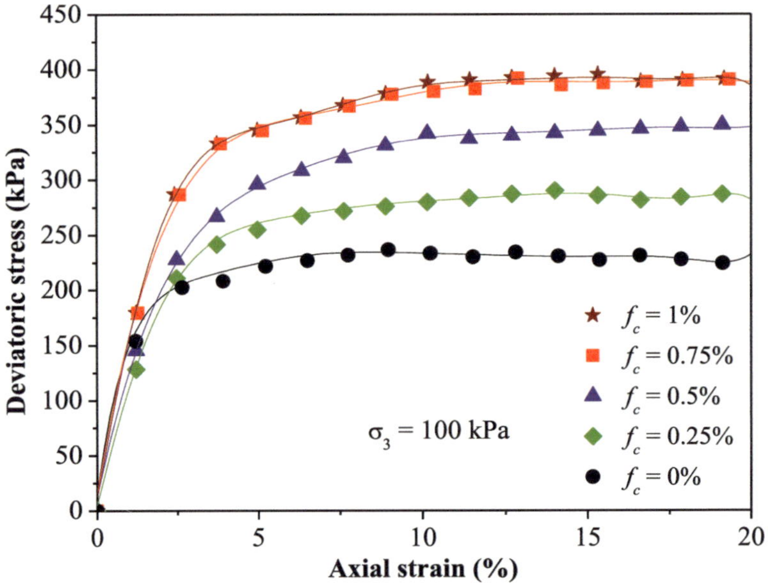

Figure 10.
Effect of fiber content on deviator stress-axial strain response.

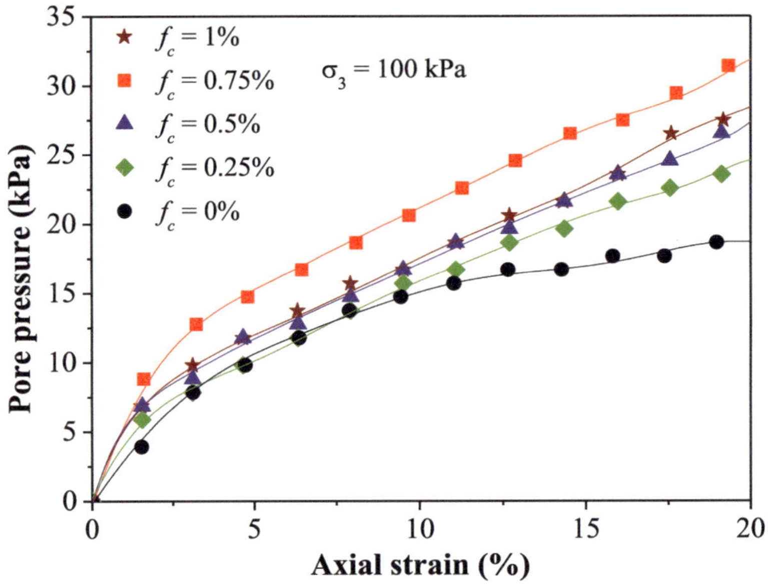

Figure 11.
Effect of fiber content on pore pressure response.

As fiber content increases, number of fiber increases within specimen which provide additional surficial friction between soil and fiber. Consequently additional mobilization of fiber tensile strength occurs with fiber content, which ultimately increases the overall strength of specimen. The initial stiffness at smaller strain (< 1%) of specimen was found to decrease with fiber content which was different from that of Ranjan et al. [35] and Estabragh et al. [22] where the initial stiffness of fiber reinforced soil was improved with fiber content. The decrease in initial stiffness with fiber content is due to the fact that the fiber within compacted specimen remains in compression at the start of shearing under confining pressure. With increasing axial strain during shearing of specimen, the fiber gets stretched by surficial interaction with soil particles and mobilizes its tensile strength resulting in improvement of strength and stiffness of the specimen.

The contraction or dilation behavior of specimen particles can be related with the generated pore water pressure during shearing and can be found by inspecting the slope of pore pressure response. The positive slope specifies the contraction behavior while negative slope indicates specimen dilation. The generated pore pressure generation was found to be positive for both unreinforced and reinforced specimens indicating contractive behavior (**Figure 11**). The positive pore water pressure generation increased with fiber content, indicating that that the increase of fiber content increased the contractive behavior of specimen by uniformly distributing the stresses within the specimen.

Stiffness is a measure of resistance offered by a material against its deformation under external applied load. Stiffness of specimen can be expressed in terms of stiffness modulus which is the ratio of stress to the corresponding axial strain. The effect of fiber content on stiffness modulus is shown in **Figure 12** under 100 kPa confining pressure. The initial stiffness of soil at smaller axial strain (<1%) is found to decrease with increasing fiber content, while at higher axial strain (> 1%) the stiffness modulus can be noted to increase with fiber content up to 0.75%. The decrease in stiffness at lower axial strain is due to the fact that reinforcement needs some stretching to mobilize its tensile strength. At smaller axial stain level as soil particles move, it try to stretch the fiber and after some deformation the fiber start to work. In this case the limiting value of that point is noted around 1%. Nevertheless, stiffness modulus remains much higher than that of unreinforced specimen with 1% fiber content.

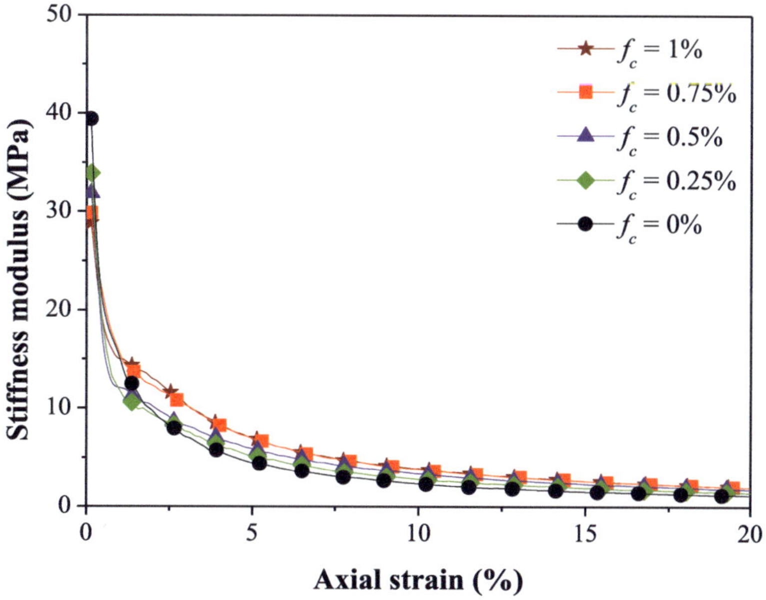

Figure 12.
Effect of fiber content on stiffness modulus response.

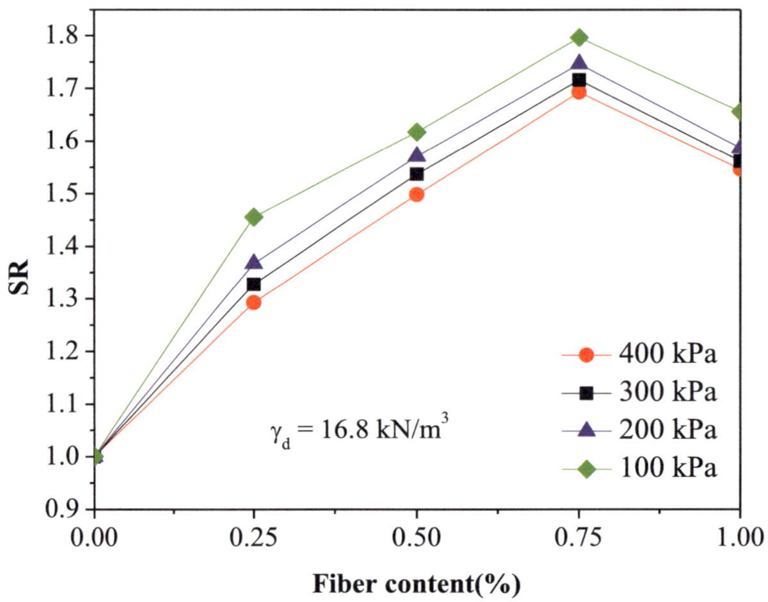

Figure 13.
Effect of fiber content on strength ratio.

For any fiber content stiffness modulus was noted to be higher at small axial strain and it progressively decreased with increasing axial strain. The stiffness modulus reduction rate decreased at higher axial strain.

Effect of fiber benefit on strength of soil during undrained shearing has been presented in terms of a parameter called strength ratio (*SR*) similar to that of Estabragh et al. [22], Haeri et al. [53] and Zhang et al. [54]. Strength ratio is the ratio of deviator stress of reinforced soil at failure (σ_{dr}) to that of deviator stress of unreinforced soil at failure (σ_{du}).

$$SR = \frac{\sigma_{dr}}{\sigma_{du}} \quad (1)$$

The influence of fiber content on *SR* under varying confinement is shown in **Figure 13**. For any fiber content, the strength ratio decreased with increasing confining pressure, indicating that the effect of fiber decreased with increasing confining pressure. It can also be noted that *SR* increased with fiber content up to 0.75% at any confining pressure and then decreased for 1% fiber content.

4. Conclusions

Following conclusions have been drawn from the experimental investigation of glass fiber-reinforced clayey soil:

1. Mixing efficiency of fibers within soil mass decreases with increasing fiber content.

2. Addition of glass fibers marginally changes the compaction parameters (OMC and MDU) of clayey soil.

3. The UCS, CBR and shear strength of clayey soil increases with glass fiber content up to a limiting value of 0.75%.

4. The addition of glass fibers enhances the UCS of clayey soil by around two fold, CBR by 2.8 times and shear strength by around 1.75 times that of unreinforced soil.

5. The glass fiber inclusion continuously increases ductility of clayey soil.

6. The inclusion of glass fibers decreases the stiffness modulus of clayey soil at smaller axial strain and then increases the stiffness at higher axial strain. The boundary of axial strain which changes the stiffness behavior is noted to be around 1%.

7. The strength ratio of clayey soil decreases with increasing confining pressure for any fiber content.

8. The 20 mm glass fibers of 0.5%, 0.75% and 1% is found to be used expressively in the subgrade layer of low-volume flexible pavement.

Author details

Suchit Kumar Patel
Central University of Jharkhand, Ranchi, India

*Address all correspondence to: suchit.patel@cuj.ac.in

References

[1] Hausmann MR. Engineering Principles of Ground Modification. Singapore: McGraw-Hill; 1990

[2] Shukla SK. Fundamentals of Fibre-Reinforced Soil Engineering. Singapore: Springer; 2017

[3] Hoover JM, Moeller DT, Pitt JM, Smith SG, Wainaina NW. Performance of Randomly Oriented fiber-Reinforced Roadway Soils. Lowa DOT Project-HR-211. Ames, United State: Department of Transportation, Highway Division, Lowa State University; 1982

[4] Waldron LJ. The shear resistance of root permeated homogeneous and stratified soil. Soil Science Society of America Journal. 1977;**41**(5):843-849

[5] Santoni RL, Webster SL. Airfields and road construction using fiber stabilization of sands. Journal of Transportation Engineering. 2001;**127**(2):96-104

[6] Rafalko SD, Brandon TL, Filz GM, Mitchell JK. Fiber reinforcement for rapid stabilization of soft clay soils. Transportation Research Record. 2026;(1):21-29

[7] Yoon S, Prezzi M, Siddiki NZ, Kim B. Construction of a test embankment using a sand-tire shred mixture as fill material. Waste Management. 2006;**26**(9):1033-1044

[8] Gregory GH, Chill DS. Stabilization of earth slope with fiber reinforcement. In: Proceedings of 6th International Conference on Geosynthetics, Atlanta; 1998. pp. 1073-1078

[9] Youwai S, Bergado DT. Numerical analysis of reinforced wall using rubber tire chips-sand mixtures as backfill material. Computers and Geotechnics. 2004;**31**(2):103-114

[10] Miller CJ, Rifai S. Fiber reinforcement for waste containment soil liners. Journal of Environmental Engineering. 2004;**130**(8):891-895

[11] Consoli NC, Prietto PDM, Ulbrich LA. Influence of fiber and cement addition on behavior of sandy soil. Journal of Geotechnical and Geoenvironmental Engineering. 1998;**124**(12):1211-1214

[12] Consoli NC, Montardo JP, Prietto PDM, Pasa GS. Engineering behavior of a sand reinforced with plastic waste. Journal of Geotechnical and Geoenvironmental Engineering. 2002;**128**(6):462-472

[13] Heineck K, Coop M, Consoli NC. Effect of microreinforcement of soils from very small to large shear strains. Journal of Geotechnical and Geoenvironmental Engineering. 2005;**131**(8):1024-1033

[14] Consoli NC, Heineck KS, Casagrande MDT, Coop MR. Shear strength behavior of fiber-reinforced sand considering triaxial tests under distinct stress paths. Journal of Geotechnical and Geoenvironmental Engineering. 2007;**133**(11):1466-1469

[15] Chen CW, Loehr JE. Undrained and drained triaxial tests of fibre-reinforced sand. In: Proceedings of 4th Asian Regional Conference on Geosynthetics. Berlin: Springer; 2008. pp. 114-120

[16] Ozkul ZH, Baykal G. Shear strength of clay with rubber fiber inclusions. Geosynthetics International. 2006;**13**(5):173-180

[17] Tang C, Shi B, Gao W, Chen F, Cai Y. Strength and mechanical behavior of short polypropylene fiber reinforced and cement stabilized clayey soil. Geotextiles and Geomembranes. 2007;**25**(3):194-202

[18] Pradhan PK, Kar RK, Naik A. Effect of random inclusion of polypropylene fibers on strength characteristics of cohesive soil. Geotechnical and Geological Engineering. 2012;**30**(1): 15-25

[19] Ozkul ZH, Baykal G. Shear behavior of compacted rubber fiber-clay composite in drained and undrained loading. Journal of Geotechnical and Geoenvironmental Engineering. 2007;**133**(7):767-781

[20] Sivakumar Babu GL, Vasudevan AK, Sayida MK. Use of coir fibers for improving the engineering properties of expansive soils. Journal of Natural Fibers. 2008;**5**(1):61-75

[21] Dasaka MS, Sumesh KS. Effect of coir fiber on the stress-strain behavior of a reconstituted fine grained soil. Journal of Natural Fibers. 2011;**8**(3):189-204

[22] Estabragh AR, Bordbar AT, Javadi AA. A study on the mechanical behavior of a fiber-clay composite with natural fiber. Geotechnical and Geological Engineering. 2013;**31**(2):501-510

[23] Freitag DR. Soil randomly reinforced with fibers. Journal of Geotechnical Engineering. 1986;**112**(8):823-826

[24] Maher MH, Ho YC. Mechanical properties of kaolinite/fiber soil composite. Journal of Geotechnical Engineering. 1994;**120**(8):1381-1393

[25] Bouhicha M, Aouissi F, Kenai S. Performance of composite soil reinforced with barley straw. Cement and Concrete Composites. 2005;**27**(5):617-621

[26] Akbulut S, Arasan S, Kalkan E. Modification of clayey soils using scrap tire rubber and synthetic fibers. Applied Clay Science. 2007;**38**(1-2):23-32

[27] Attom MF, Al-Akhras NM, Malkawi AIH. Effect of fibres on the mechanical properties of clayey soil. Proceedings of the Institution of Civil Engineers: Geotechnical Engineering. 2009;**162**(5):277-282

[28] Mirzababaei M, Miraftab M, Mohamed M, McMahon P. Unconfined compression strength of reinforced clays with carpet waste fibers. Journal of Geotechnical and Geoenvironmental Engineering. 2013;**139**(3):483-493

[29] Patel SK, Singh B. Strength and deformation behavior of fiber-reinforced cohesive soil under varying moisture and compaction states. Geotechnical and Geological Engineering. 2017;**35**(4):1767-1781

[30] Divya PV, Viswanadham BVS, Gourc JP. Evaluation of tensile strength-strain characteristics of fiber-reinforced soil through laboratory tests. Journal of Materials in Civil Engineering. 2014;**26**(1):14-23

[31] Tang CS, Shi B, Zhao LZ. Interfacial shear strength of fiber reinforced soil. Geotextiles and Geomembranes. 2010;**28**(1):54-62

[32] Chandra S, Viladkar MN, Nagrale PP. Mechanistic approach for fiber-reinforced flexible pavements. Journal of Transportation Engineering. 2008;**134**(1):15-23

[33] Edincliler A, Cagatay A. Weak subgrade improvement with rubber fiber inclusions. Geosynthetics International. 2013;**20**(1):39-46

[34] Patel SK, Singh B. Experimental investigation on the behavior of glass

fibre-reinforced cohesive soil for application as pavement subgrade material. International Journal of Geosynthetics and Ground Engineering. 2017;**3**(2):1-12

[35] Ranjan G, Vasan RM, Charan HD. Probabilistic analysis of randomly distributed fiber-reinforced soil. Journal of Geotechnical Engineering. 1996;**122**(6):419-426

[36] Al-Akhras NM, Attom MF, Al-Akhras KM, Malkawi AIH. Influence of fibres on swelling properties of clayey soil. Geosynthetics International. 2008;**15**(4):304-309

[37] Tang C, Shi B, Cui Y, Liu C, Gu K. Desiccation cracking behaviour of polypropylene fibre-reinforced clayey soil. Canadian Geotechnical Journal. 2012;**49**(9):1088-1101

[38] Viswanadham BVS, Phanikumar BR, Mukherjee RV. Swelling behaviour of a geofibre-reinforced expansive soil. Geotextiles and Geomembranes. 2009;**27**(1):73-76

[39] Ang EC, Loehr JE. Specimen size effects for Fiber-reinforced silty clay in unconfined compression. Geotechnical Testing Journal. 2003;**26**(2):191-200

[40] ASTM D 2487. Standard Practice for Classification of Soils for Engineering Purposes (Unified Soil Classification System). West Conshohocken, PA: ASTM International; 2006

[41] ASTM D 698. Standard Test Methods for Laboratory Compaction Characteristics of Soil Using Standard Effort (12,400 Ft-Lbf/ft^3 (600 kN-M/m^3)). West Conshohocken, PA, USA: ASTM International; 2012

[42] Lutz JT, Grossman RF. Polymer modifiers and additives. In: Lopez-Anido RA, Naik TR, Fry GT, Lange DA, Karbhari VM, editors. Emerging Material for Civil Infrastructure: State of the Art. Reston, VA: ASCE; 2001

[43] Mujah D, Ahmad F, Hazarika H, Safari A. Evaluation of the mechanical properties of recycled glass Fibers-derived three dimensional Geomaterial for ground improvement. Journal of Cleaner Production. 2013;**52**:495-503

[44] Ahmad F, Mujah D, Hazarika H, Safari A. Assessing the potential reuse of recycled glass fibre in problematic soil applications. Journal of Cleaner Production. 2012;**35**:102-107

[45] ASTM D 2166/D 2166M. Standard Test Method for Unconfined Compressive Strength of Cohesive Soil. West Conshohocken, PA, USA: ASTM International; 2013

[46] ASTM D4767. Standard Test Method for Consolidated Undrainedtriaxial Compression Test for Cohesive Soils. West Conshohocken: ASTM International; 2011

[47] ASTM D 1883. Standard Test Method for California Bearing Ratio (CBR) of Laboratory-Compacted Soils. West Conshohocken, PA, USA: ASTM International; 2016

[48] Falorca IMCFG, Pinto MIM. Effect of short, randomly distributed polypropylene microfibres on shear strength behaviour of soils. Geosynthetics International. 2011;**18**(1):2-11

[49] Al-Refeai TO, Al-Suhaibani A. Dynamic and static characterization of polypropylene fiber-reinforced dune sand. Geosynthetics International. 1998;**5**(5):443-458

[50] IRC SP 72. Guidelines for the Design of Flexible Pavements for Low Volume

Rural Roads. New Delhi, India: The Indian Road Congress; 2007

[51] IRC 37. Guidelines for the Design of Flexible Pavements. New Delhi, India: The Indian Road Congress; 2001

[52] Andersland OB, Khattak AS. Shear strength of kaolinite/fiber soil mixture. In: Proceedings of 1st International Conference on Soil Reinforcement, Paris. Vol. 1. 1979. pp. 11-16

[53] Haeri SM, Noorzad R, Oskoorouchi AM. Effect of geotextile reinforcement on the mechanical behavior of sand. Geotextiles and Geomembranes. 2000;**18**(6):385-402

[54] Zhang MX, Javadib AA, Min X. Triaxial tests of sand reinforced with 3D inclusions. Geotextiles and Geomembranes. 2006;**24**(4):201-209

Section 2

Foundation Analysis and Design

Chapter 4

Reducing Carbon Emissions by Combined Pile-Raft Foundations for High-Rise Structures

Rolf Katzenbach and Steffen Leppla

Abstract

Regarding the impact of construction processes on the environment, the reduction of CO_2 has an important role. The production of materials e.g. reinforced concrete, and the construction of structures consume large amounts of energy, which leads to a large emission of CO_2. The target is the reduction of the amount of construction material used and of the energy consumed for construction. For this, the structures have to be optimized regarding the geometry considering the requirements of the stability, serviceability, and durability. Also, foundation systems of high-rise buildings can be optimized regarding CO_2 emission. For the optimization, three parts have to be considered. The first part is the detection of the real load-deformation behavior of a foundation element. This can be reached by large-scale load tests in situ. The second part is to use the hybrid foundation system Combined Pile-Raft Foundation (CPRF), which combines the bearing capacities of the raft and of the piles. The third part is the realistic prediction of the load-deformation behavior of the foundation. For this three-dimensional, nonlinear calculations using the Finite-Element-Method (FEM) are necessary. The contribution explains the three parts and shows the application in engineering praxis, including case studies.

Keywords: CO_2 reduction, load test, Combined Pile-Raft Foundation, high-rise building, sustainability

1. Introduction

The most important aspects for the design of any foundation system are safety, serviceability, and sustainability. The requirements for safety and serviceability are defined in standards, codes, and regulations. For sustainable construction, a reduction of construction material used and energy consumed during the construction phase and the service phase of a building/structure is important. Regarding the changing climate and the necessity to avoid CO_2 emissions, the design and construction of new buildings and structures have to be optimized. The focus has to be on the production of cement. The production of one ton of cement leads to an emission of about 800 kg of CO_2. This is about 91% of the whole CO_2 footprint of concrete and about 8% of the man-made CO_2 emission of the world [1]. This shows that the reduction of concrete for any kind of structure is an important aspect for the reduction of CO_2 emission.

Optimized foundations systems lead to a reduction of concrete. This optimization has to consider the requirements of safety, serviceability, and sustainability. For the foundation systems of high-rise buildings, the following parts are necessary:

- Large-scale load tests in situ on the construction site to detect the real load-deformation behavior of the foundation.
- Hybrid foundation systems for high-rise buildings like the Combined Pile-Raft Foundation (CPRF) [2].
- Three-dimensional, nonlinear simulations of the load-deformation behavior of the foundation system using e.g. Finite-Element-Method (FEM).

All of these three important aspects will be explained in the following chapter. Nevertheless, the precondition for any kind of safe and optimized design is a sufficient soil and groundwater investigation.

2. Large scale in situ load tests of piles

Load tests of piles, that are performed in-situ on the construction site are the best opportunity for the determination of the load-deformation behavior [3]. For the determination of the bearing capacity, the loads on test piles can have a vertical resp. horizontal direction. Vertical loads can be compression loads or tension loads depending on the construction task. The tests can be static load tests or dynamic load tests. Detailed descriptions of these different test types are given in [4, 5]. In the following, only the static pile load test for determining the vertical bearing capacity is presented.

Normally counterweights or anchors are used as an abutment for the pile load. The installation of counterweights or anchors necessitates large technical and financial input. Using hydraulic jacks like the Osterberg-cell (O-cell) is more convenient. **Figure 1** shows the variations of static pile load tests.

By using the Osterberg-method, hydraulic jacks are installed in a test pile to detect to determine the skin friction in different pile segments that correspond to different soil layers. The single pile segments serve as counterweights for the different test phases.

The result of a pile load test with vertical load is described by a resistance settlement curve $R_{c,k}(s)$ which can be used as the basis for the analyses of stability and serviceability. In **Figure 2** a qualitative trend of a resistance settlement curve is shown. Two straight reference lines help to determine the pile resistance $R_{c,k}$. These two straight reference lines draw a tangent at the beginning and at the end of the resistance settlement curve. The interaction of both lines defines the stability limit state.

Based on one or several pile load tests, the measured value $R_{c,m}$ is determined, which has to be reduced by the factor ξ taking straggling into account. According to [6] the pile resistance has to be calculated by Eq. (1) if the superstructure is not able to transfer loads from softer to stiffer piles.

$$R_{c,k} = MIN\left\{\frac{(R_{c,m})_{av}}{\xi_1};\frac{(R_{c,m})_{min}}{\xi_2}\right\} \tag{1}$$

The superstructure is able to transfer loads from softer to stiffer piles if the superstructure has sufficient rigidity. In this case, the straggling factors ξ_i can be

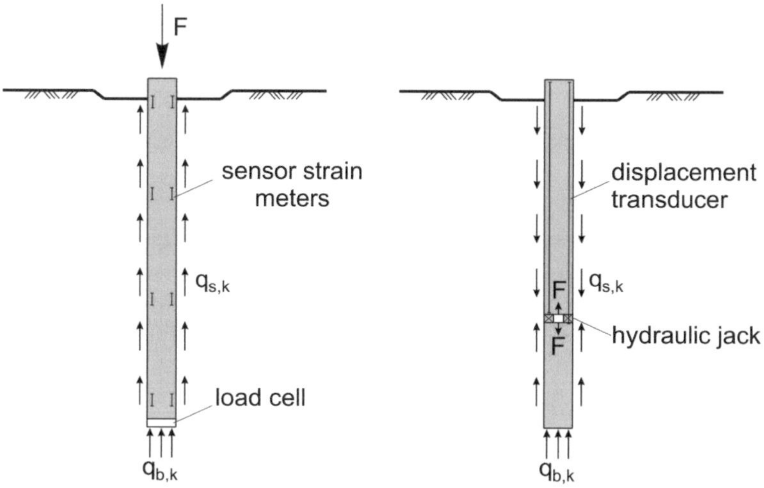

Figure 1.
Static piles load test with counterweight resp. anchor (left) and hydraulic jack (right).

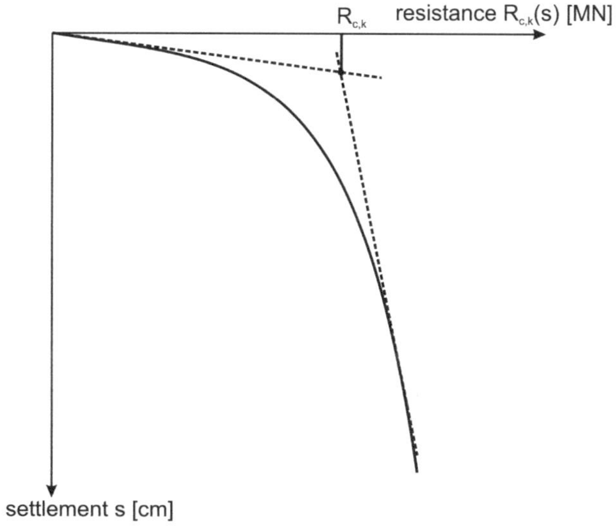

Figure 2.
Determination of the pile resistance by a resistance settlement curve.

n	1	2	3	4	≥5
ξ_1	1.35	1.25	1.15	1.05	1.00
ξ_2	1.35	1.15	1.00	1.00	1.00
n = number of pile load tests					

Table 1.
Straggling factors ξ_i for resistance of pressure piles.

divided by 1.1 (ξ_1 is always ≥ 1.0). To the measured average pile resistance belongs the straggling factor ξ_1. To the measured minimum pile resistance belongs the straggling factor ξ_2. The straggling factors for pressure piles are given in **Table 1**.

3. Combined Pile-Raft Foundation (CPRF)

3.1 Basics

A Combined Pile-Raft Foundation (CPRF) is a hybrid, technically and economically optimized foundation system. It combines the bearing capacity of a foundation raft and of piles or barrettes. For the foundation of classic high-rise buildings as well as for engineering constructions like bridges and towers CPRFs can be used.

The technical regulations for classic deep foundations prevail for CPRFs as well [4]. In addition, the Combined Pile-Raft Foundation Guideline [7] has to be considered. This internationally validated guideline reflects the individual features of a CPRF and is published by the International Society for Soil Mechanics and Geotechnical Engineering (ISSMGE).

CPRFs have a very complex bearing and deformation behavior due to the interaction between the foundation elements and the subsoil. CPRFs belong to the Geotechnical Category GC 3 according to EC 7 [6].

The advantages of a CPRF, compared to a conventional spread foundation and a classic pile foundation, are the reduction of:

- Settlements and differential settlements.
- The bending moments of the foundation raft.
- Pile materials (30–40%)

3.2 Bearing and deformation behavior

The measurement data of high-rise buildings founded on spread foundations in Frankfurt am Main, Germany, showed, that 60–80% of the settlements arise in the upper third of the influenced soil volume. A part of the load on a CPRF is transferred py the piles from areas with a small stiffness under the foundation raft to a stiffer, deeper area of the subsoil without neglecting the bearing capacity of the foundation raft (**Figure 3**).

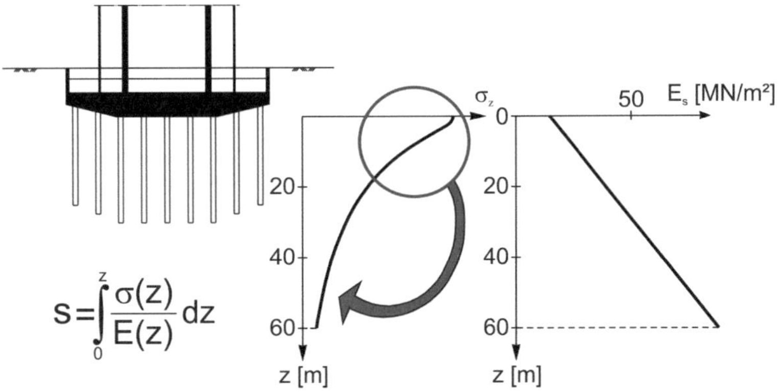

Figure 3.
Principle load transfer of a CPRF.

The bearing and deformation behavior of a CPRF is characterized by the interaction between the bearing elements (foundation raft and pile resp. barrettes) and the subsoil. **Figure 4** shows all interactions of a CPRF.

A CPRF transfers the total building load $F_{tot,k}$ to the piles and the subsoil. The mobilized resistance of a CPRF depends significantly on the settlement s, which is similar to a classic deep foundation. The resistance $R_{raft,k}(s)$ equates to the integration of the soil contact pressure $\sigma(x,y)$ under the foundation raft. The resistance $R_{tot,k}(s)$ of a CPRF equates to the resistance of the foundation piles $\sum R_{pile,k,i}(s)$ added to the resistance of the foundation raft $R_{raft,k}(s)$ (Eq. (2)).

$$R_{tot,k}(s) = \sum_{i=1}^{i=n} R_{pile,k,i}(s) + R_{raft,k}(s) \quad (2)$$

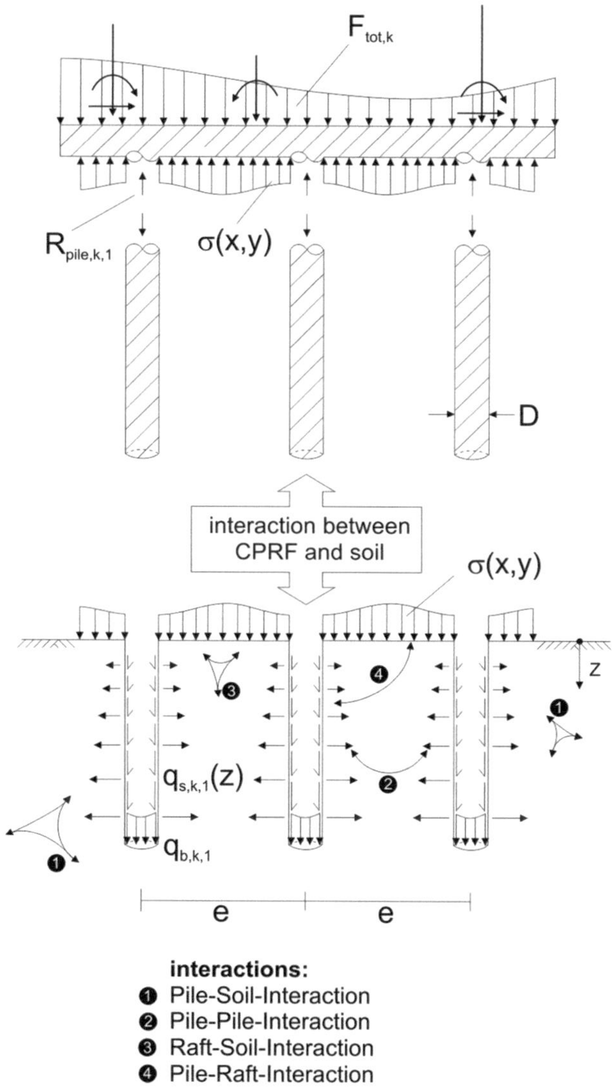

Figure 4.
Interactions of a CPRF.

As shown in Eq. (3), the total resistance of a single foundation pile consists of the skin resistance $R_{s,k,i}(s)$ and the pile base resistance $R_{b,k,i}(s)$. The skin resistance $R_{s,k,i}(s)$ can be calculated by integration of the skin friction $q_{s,k}(s,z)$, which depends on the settlement s and the depth z.

$$\begin{aligned} R_{pile,k,i}(s) &= R_{b,k,i}(s) + R_{s,k,i}(s) \\ &= q_{b,k,i} \bullet \frac{\pi \bullet D^2}{4} + \int q_{s,k,i}(s,z) \bullet \pi \bullet D \bullet dz \end{aligned} \tag{3}$$

The load-deformation behavior of a CPRF can be specified by the CPRF coefficient α_{CPRF}. This coefficient declares the relation between the resistance of the piles and the total resistance and varies between 0 and 1 (Eq. (4)).

$$\alpha_{CPRF} = \frac{\sum R_{pile,k,i}(s)}{R_{tot,k}(s)} \tag{4}$$

If the whole load $F_{tot,k}$ is carried by the foundation raft, the CPRF coefficient is $\alpha_{CPRF} = 0$. If the whole load $F_{tot,k}$ is carried by the foundation piles, the CPRF coefficient is $\alpha_{CPRF} = 1$. Related to technical and economic aspects a CPRF coefficient α_{CPRF} between 0.5 and 0.7 can be considered as optimum. For $\alpha_{CPRF} > 0.9$ additional analyses on the piles are necessary.

The effective horizontal stresses influence the mobilized skin friction of the piles. Hence the stress level of the subsoil influences the load-deformation behavior of a CPRF. The neighboring piles, the foundation raft, and the effects during the construction of the piles influence the stress level of the subsoil around every pile of a CPRF. The soil contact pressure under the foundation raft leads to an increased stress level of the subsoil. The result is higher skin friction in the upper parts of the piles.

3.3 Principle calculation method of a CPRF

For the design and calculation of a CPRF various methods can be selected [8–14]. Up to now only numerical methods, like the Finite-Element-Method (FEM) provide calculation results that are comparable to reality.

The knowledge about the load-deformation behavior of a free, single pile is necessary for a qualified design of a CPRF [4]. Otherwise, a pile load test has to be performed. Two reasons are important for the knowledge about the bearing capacity of a free, single pile:

- Evaluation of the selected geometries of the piles and to prove the plausibility of the calculation method.

- Possibility to calibrate the numerical model.

In situ pile load tests are required for complex construction projects and/or difficult soil conditions.

3.4 Monitoring of a CPRF

Regarding the Geotechnical Category GC 3 a CPRF has to be monitored [4, 6, 7]. The monitoring program consists of geodetic and geotechnical measurements of the new building and of the vicinity and covers the construction phase and the service phase of the building. The following tasks are important:

- Verification of the calculation model including the parameters used.
- Early detection of critical forces, stresses, deformations.
- Verification of the predicted deformations.
- Quality assurance and preservation of evidence.

4. Examples from engineering practice

4.1 Calibration of a numerical model

Numerical simulations using FEM have been carried out for the design of a CPRF of a new high-rise building founded in soft soil [15]. For the calibration of the numerical model a pile load test using Osterberg-Cells (O-cells) has been carried out in the project area. The test pile consisted of the upper test segment 1, the middle test segment 2 between the two O-cells, and the lower test segment 3.

In various testing phases, the O-cells were activated individually to determine the skin friction of the different layers and the pile base resistance. At test segment 3 only the lower O-cell was activated, while test segment 2 was used as an abutment to determine the skin friction and the pile base resistance. At test segment 2 the upper O-cell was activated and the lower O-cell was released to determine the skin friction. Test segment 1 was used as an abutment in this test phase. At test segment 1 the upper O-cell was activated and the lower O-cell was stiffened to determine the skin friction. Test segments 2 and 3 were used as an abutment in this test phase.

A numerical (FEM) back analysis of the pile load test was used to calibrate the numerical model of the CPRF. The FE-model of the numerical back analysis of the pile load test with the three test segments and the two O-cells is shown in **Figure 5**.

The results of the in situ pile load test and the numerical back analysis show good accordance (**Figure 6**). By this, the used soil mechanical parameters and the simplified stratigraphy, which was necessary for the numerical model, were verified.

The design of the CPRF is performed by three-dimensional, nonlinear FE-simulations. Taking into account the requirements of the load-deformation behavior the length, diameter, and the number of piles were optimized on the basis of the FE-simulations. The optimized CPRF is shown in **Figure 7**. Eighty percent of the total building load are carried by the piles and 20% of the total building load is carried by the raft. So, the CPRF coefficient is $\alpha_{CPRF} = 0.8$.

4.2 High-rise building in settlement active clay

The high-rise building Messeturm in Frankfurt am Main, Germany, is 256.5 m high and is founded on a CPRF in the settlement active Frankfurt Clay (**Figure 8**). The foundation raft has a ground view of 58.8 m × 58.8 m with a maximum thickness of 6 m in the center and a thickness of 3 m at the edges. The base of the foundation raft is about 11–14 m below the surface.

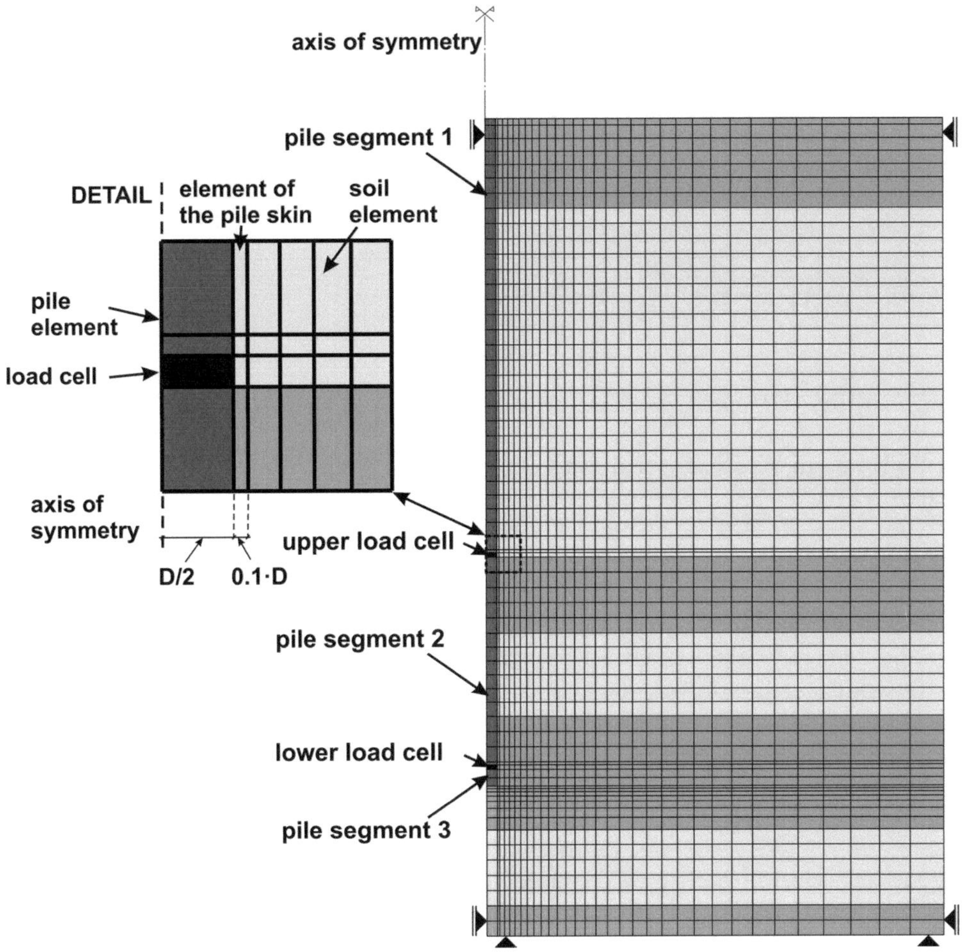

Figure 5.
Numerical simulation of the pile load test for calibration.

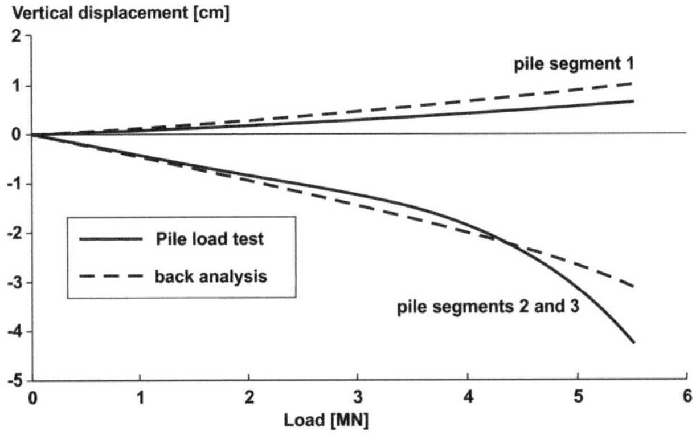

Figure 6.
Measurement and calculation of the pile load test (upper O-cell activated, lower O-cell stiffened).

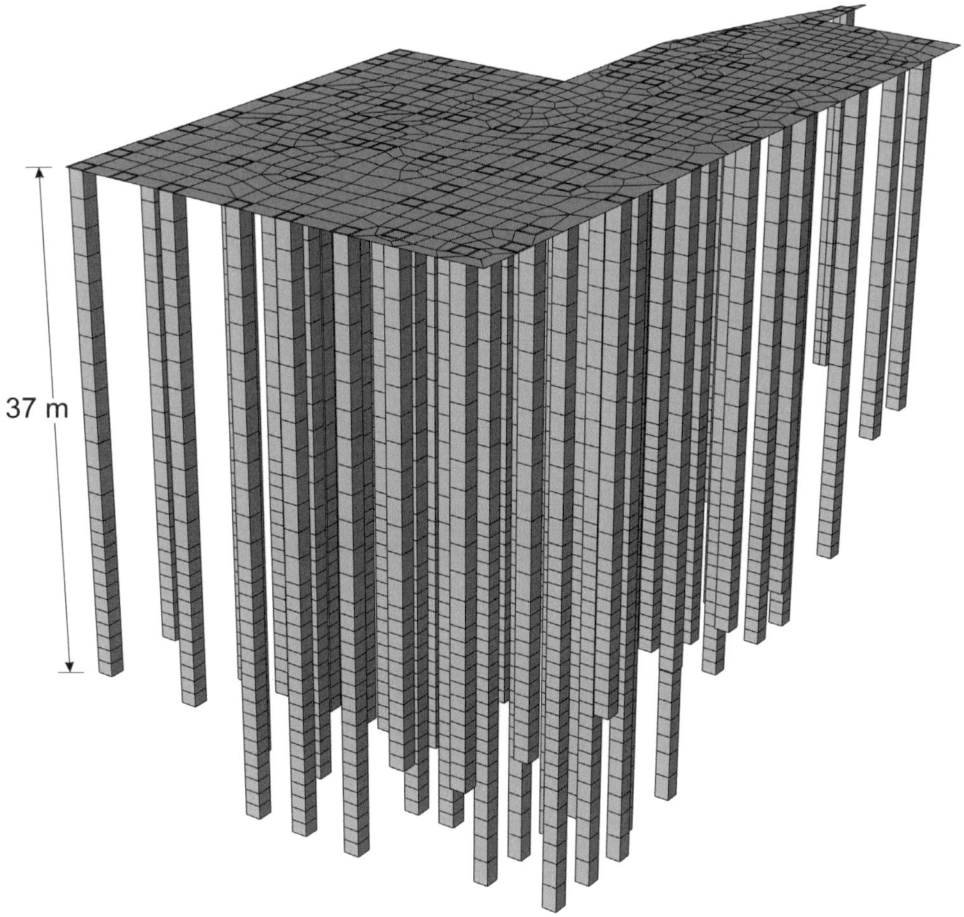

Figure 7.
FE-mesh of the optimized CPRF.

Figure 9 shows the CPRF with 64 bored piles with a diameter of 1.3 m. The length varies between 30.9 m in the center and 26.9 m at the edges. The total building load is about 1,855 MN including 30% of the live loads.

The subsoil in the project area consists of artificial fillings at the surface which are underlain by quaternary sand and gravel until a depth of 8–10 m below the surface. Below follows the tertiary Frankfurt Clay to a depth of about 70 m below the surface. At a depth of 4.5–5.0 m below the surface is the groundwater table. The maximum measured settlements of the foundation raft were 13 cm in the center and 7–9 cm at the edges.

The CPRF was calculated using the FEM. Thereby a section of the foundation was modeled, using the symmetry of the ground view (**Figure 10**).

The FE-calculation simulates the construction process step-by-step. These steps are the excavation of the construction pit, the construction of the CPRF, the groundwater lowering, the loading of the CPRF, and the groundwater re-increase.

For the optimization of the CPRF different pile configurations and pile length was analyzed as well as a pure raft foundation. **Figure 11** shows the comparison of the load-settlement curves of a pure raft foundation and of a CPRF.

Figure 8.
High-rise building Messeturm in Frankfurt am Main, Germany.

The maximum settlements of a pure raft foundation were calculated to be 32.5 cm. The in situ measured maximum settlements of the CPRF of 13 cm correspond to the calculated maximum settlements. The calculation and the measurement data showed a CPRF coefficient of $\alpha_{CPRF} = 0.43$.

Until the construction of the Messeturm the ultimate skin friction q_s of bored piles in Frankfurt Clay was estimated to 60–80 kN/m^2 for 20 m long piles, based on pile load tests. At the piles of the Messeturm, an average skin friction q_s of 90–105 kN/m^2 was measured. At the pile toe, a maximum skin friction q_s of 200 kN/m^2 was measured.

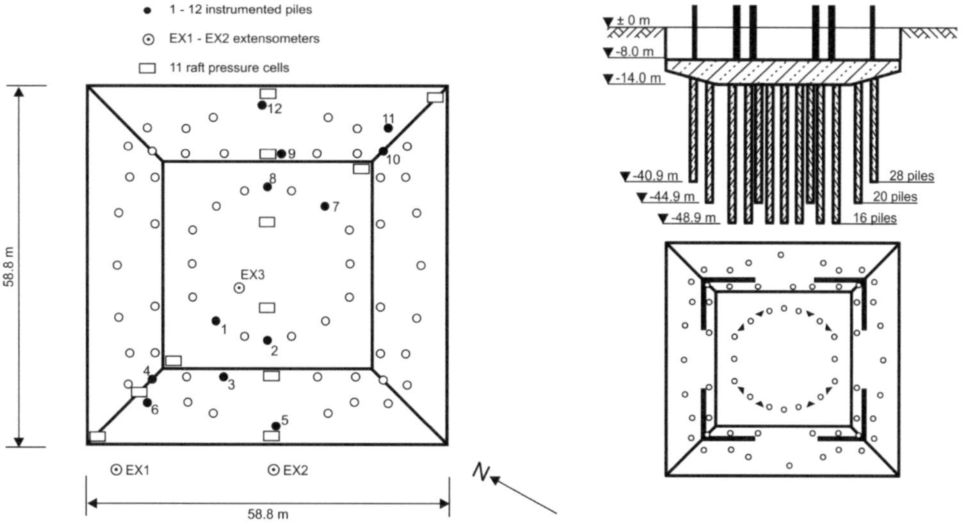

Figure 9.
Ground view (left) and cross-section (right) of the CPRF.

A pure pile foundation would have required 316 piles with 30 m in length and a diameter of 1.3 m. In comparison to the realized CPRF with 64 piles and an average length of 30 m, a pure pile foundation would have required much more material, time, and money. Regarding the CO_2 emission, the CPRF saved about 10,000 tons of concrete. With the estimation, that the average cement ratio is about 300 kg/t of concrete, the CPRF saved about 6000 t of CO_2.

4.3 High-rise building on a steep slope

The high-rise building Mirax Plaza in Kiev, Ukraine, consists of two high-rise buildings, each of them with a height of 192 m (**Figure 12**). The subsoil consists of artificial fillings to a depth of 2–3 m, which are underlain by quaternary silty sand and sandy silt with a thickness of 5–10 m. Below follow tertiary silt and sand with a thickness of 0–24 m. Then follows tertiary clayey silt and clay marl of the Kiev and Butschak formation with a thickness of about 20 m, which is underlain by tertiary fine sands of the Butschak formation. The groundwater level is about 2 m under the service. The soil conditions and a cross-section of the construction project are shown in **Figure 13**.

Two pile load tests have been carried out on the construction site to verify the skin and the base resistance of the deep foundation elements and for the calibration of the numerical simulations. The piles had a length of 10 m and 44 m and a diameter of 0.82 m. The soil properties that resulted from the back analysis were partly three times higher than indicated in the geotechnical report. The results of the numerical back analysis and the load tests show good accordance (**Figure 14**).

Tower A has a foundation raft of about 2000 m^2 and an overall load of about 2200 MN. **Figure 15** shows the calculated settlements of the three-dimensional FEM simulation.

The raft is located at a depth of 10 m below the surface in Kiev clay marl. The barrettes go through the Kiev clay marl and reach the tertiary fine sands.

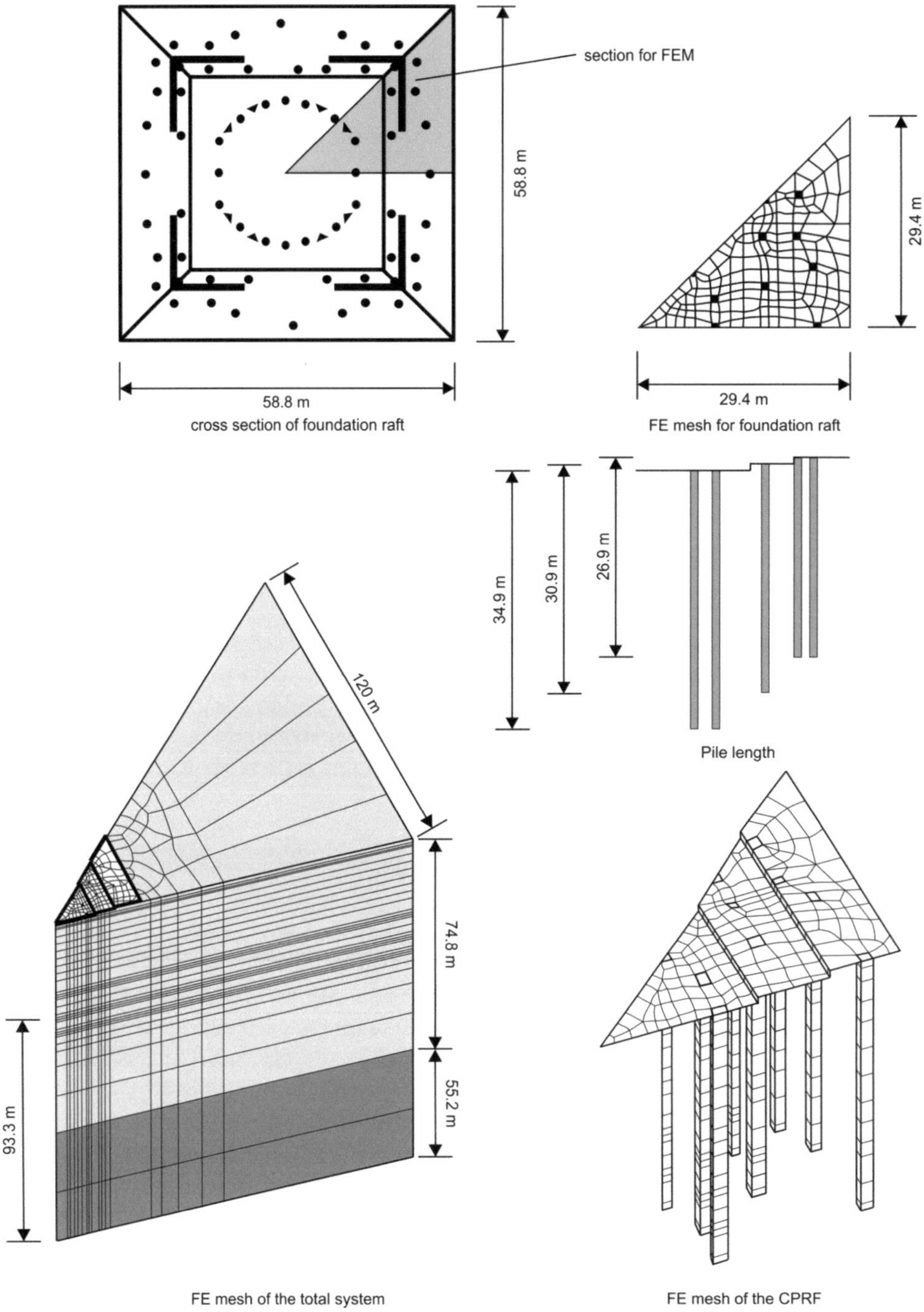

Figure 10.
FE-mesh of numerical simulation.

The outer barrettes have calculated loads between 41.2 MN and 44.5 MN. The inner barrettes have calculated load between 22.1 MN and 30.7 MN. This is typical behavior of a CPRF. The barrettes at the edge of the foundation raft have a higher

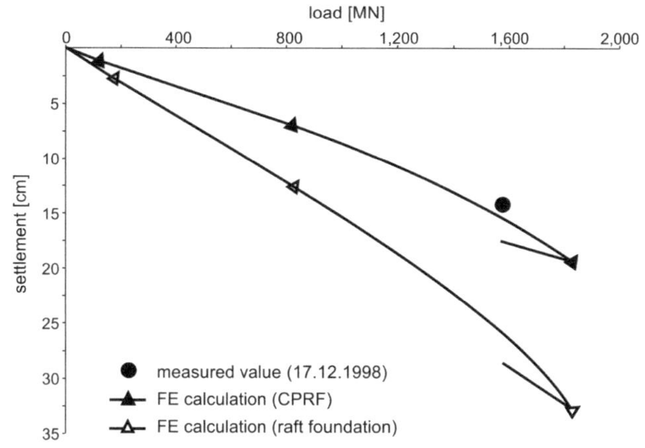

Figure 11.
Measured and calculated settlements.

Figure 12.
Mirax Plaza in Kiev, Ukraine.

stiffness due to the bigger volume of the activated soil. They get more of the total load. The calculated CPRF coefficient is α_{CPRF} = 0.88. The settlement-relevant load of 85% of the total load will lead to maximum settlements of about 12 cm. The estimated pressure under the raft is about 200 kN/m^2 (center) and 400 kN/m^2 (edges).

The calculated base pressure under the barrettes is about 4130 MN/m^2 (center) and 5100 MN/m^2 (edges). The estimated skin friction increases with the depth reaching 150 MN/m^2 (center) to 180 kN/m^2 (edges).

The foundation of Mirax Plaza is the first authorized CPRF in Ukraine. The CPRF reduced the number of barrettes from 120 with 40 m length to 64 with 33 m length. Regarding the CO_2 emission, the CPRF saved about 15,000 tons of concrete. With the

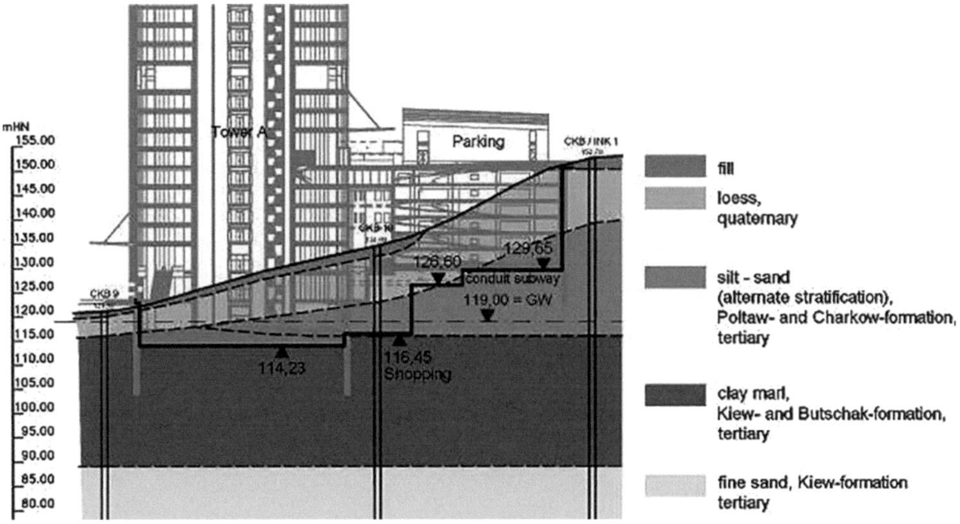

Figure 13.
Soil conditions and cross-section of Mirax Plaza.

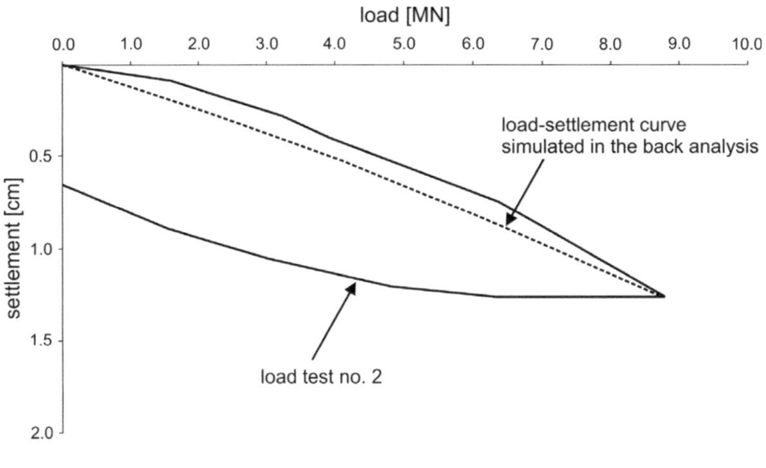

Figure 14.
Measurement and calculation of the in situ load test.

estimation, that the average cement ratio is about 300 kg/t of concrete, the CPRF saved about 9000 t of CO_2.

4.4 Settlement sensitive structure on a geological fault

The soil investigation for the science and congress center Darmstadtium in Darmstadt, Germany, showed that the planned settlement-sensitive structure is situated above the eastern fault of the Rhine Valley. The construction was finished in 2007 and is shown in **Figure 16**.

The eastern fault of the Rhine valley crosses the project area as shown in **Figure 17**. In the northern and western areas unconsolidated sediments of the Rhine Valley fault were found. In the eastern and southern area, rocks of the Odenwald crystalline were identified (granodiorite). The tectonic activities along the fault zone have not finished

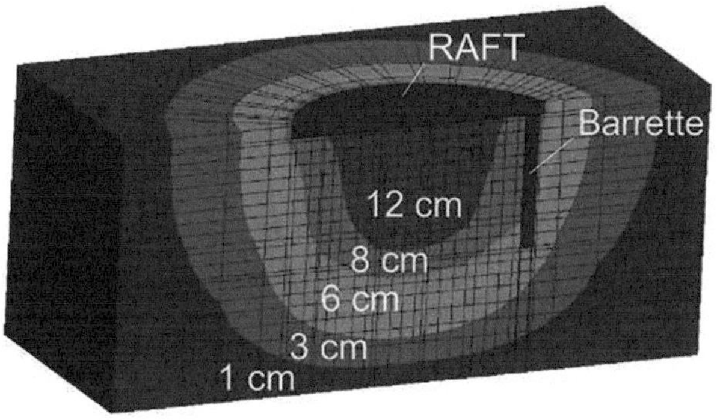

Figure 15.
FE-mesh of the numerical model and calculated settlements.

Figure 16.
Science and congress center Darmstadtium, Germany.

up to now. The area of Darmstadt that is located west of the Rhine Valley fault has an annual settlement of about 0.5 mm. These tectonic displacements hand to be considered for the design of the foundation system and the rising structure. In the area of the rock, the foundation was constructed as a spread foundation and a CPRF was constructed in the area of the Rhine Valley (**Figure 18**).

4.5 Horizontal loads on a CPRF

The Exhibition Hall 3 in Frankfurt am Main, Germany, was finished in 2001 and is one of the biggest exhibition halls in Europe. Its length is about 210 m and its width is about 130 m. The height is about 45 m. The roof is a double-curved, three-dimensional, load-bearing structure consisting of five arched compression trusses and

Figure 17.
Excavation pit and gradient of fault.

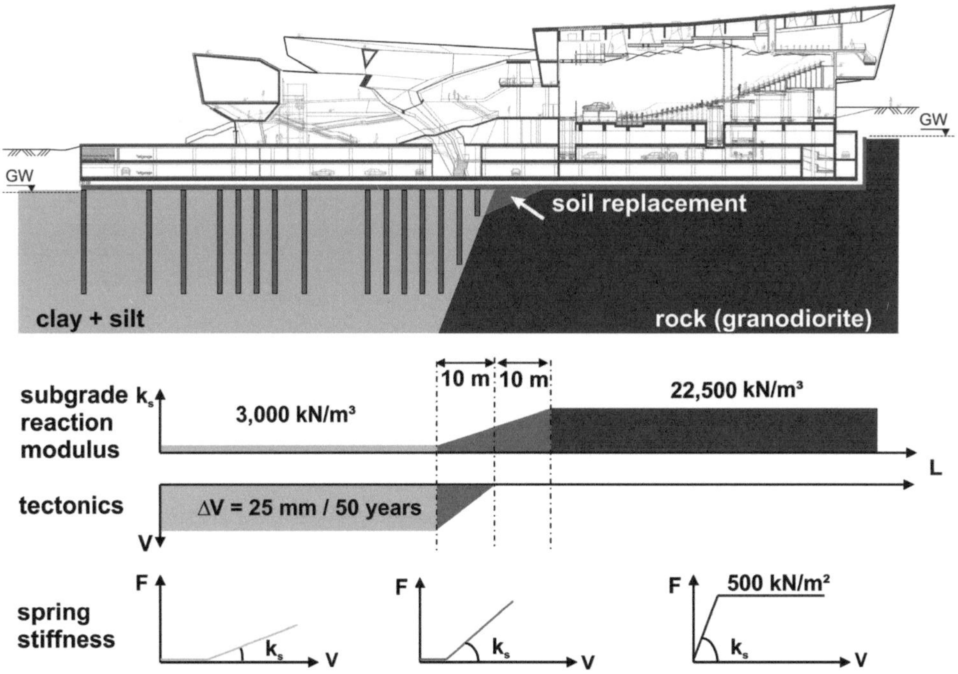

Figures 18.
Foundation system.

six arched tension trusses with a free span of 165 m [16, 17]. **Figure 19** shows a cross-section of the realized project and the subsoil conditions. Twelve A-frames, six on each side, carry the horizontal and vertical loads of the roof. These A-frames, with a height of 24 m, are constructed of two steel tubes (**Figure 20**). According to [6] the project belongs to the Geotechnical Category GC 3.

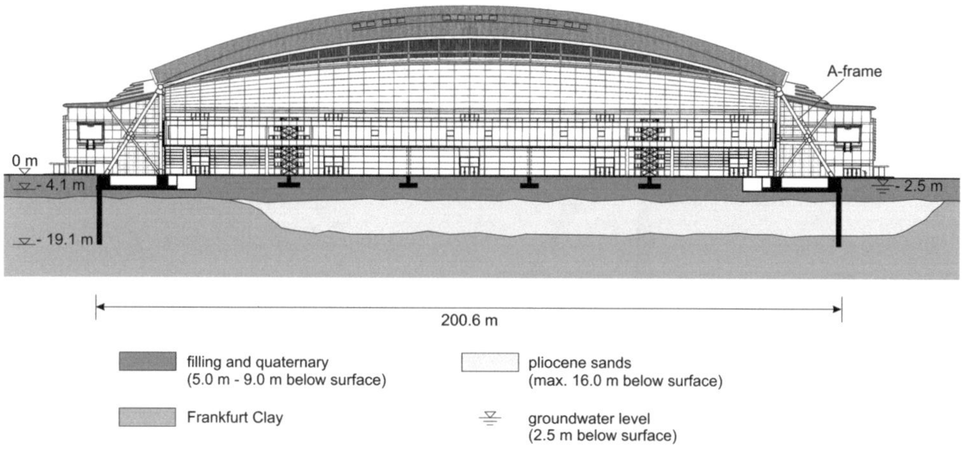

Figure 19.
Cross-section and subsoil conditions.

The soil investigation showed, that the conditions are not equal all over the project area. Under the surface is a 5–9 m thick layer of fillings and quaternary soil. Below this follows a layer of tertiary sediments. The project area is crossed diagonally by a layer of tertiary sand and gravel. The settlement active Frankfurt Clay follows until bigger depth.

A strong limitation of the displacements of the foundation is necessary due to the strong interaction between the superstructure, the foundation, and the subsoil. Three-dimensional numerical analyses were used for the design of the horizontal loaded CPRF. On each end of the hall is a CPRF which consists of a raft and 14 bored piles. The raft has a thickness of 1.4 m, a length of 127.5 m, and a width of 22.15 m. The bored piles have a diameter of 1.5 m and a length of 15 m.

According to the observational method, a geotechnical and a geodetic measurement program was installed. By four inclinometers the horizontal displacements of the CPRF were observed at a depth of 50 m under the surface. For the measurement of the vertical displacements, four extensometers were installed. In addition. pressure cells in the soil under the raft, strain gauges at A-frames, and geodetic measurement points were installed. The measurements showed horizontal displacement up to 1 cm and vertical displacements between 1.0 cm and 3.5 cm.

The example shows that the CPRF can be used for a settlement-reduced transfer of horizontal loads into the subsoil. Compared to a classic file foundation or a massive block foundation the CO_2 emission was reduced significantly.

4.6 High-rise building on cavernous subsoil conditions

The project Moscow City contains several high-rise building for business in Moscow, Russia, on an area of more than one square kilometer [18]. In this project, the Federation Tower is a complex of two single towers (**Figure 21**). Tower A is about 374 m high, or 450 m high when including the spire on the roof. The height of Tower B is about 243 m. At the start of the construction in 2003, the high-rise double-towers were planned as the highest high-rise buildings in Russia and Europe. The two towers are founded on a foundation raft, which is 4.6 m thick and has a length of 140 m and a width of 80 m. The foundation level is about 20 m below the surface.

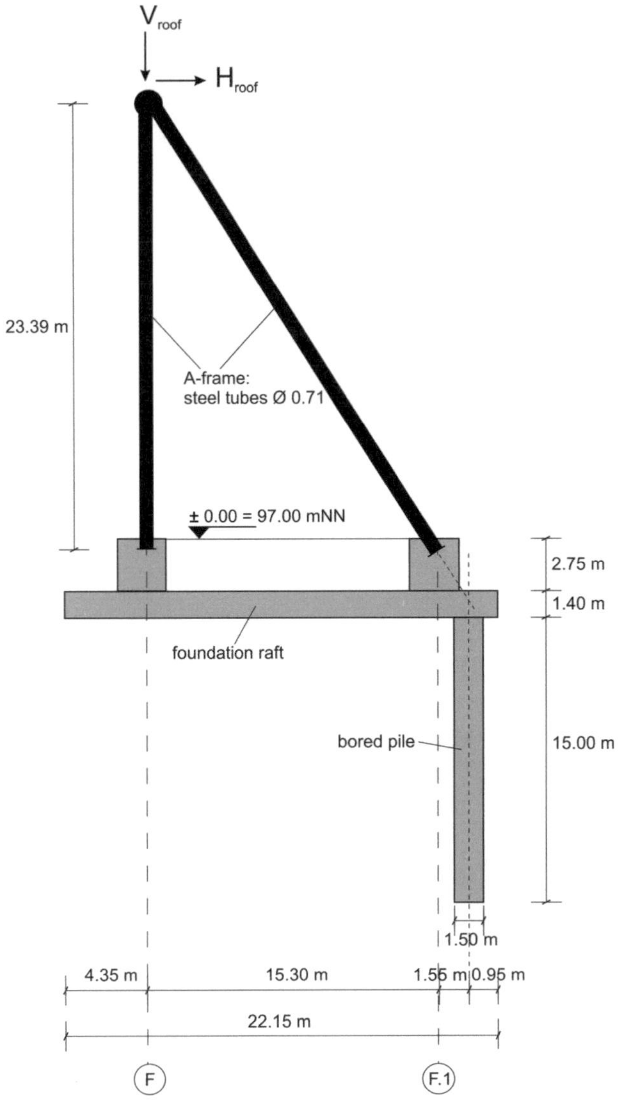

Figure 20.
Schematic visualization of the A-frames and the CPRF.

Tower A has a total load of about 3000 MN and Tower B has a total load of about 2000 MN. Including loads of about 1000 MN for adjacent buildings and the basement floors and a load of about 1300 MN for the foundation raft itself, the total load results in 7300 MN.

The project area of Moscow City is located on the left bank of the River Moskva in the west of the central district. The anthropogenic artificial fillings are followed by the quaternary accumulation of the river terrace. Below this, an alternating sequence of carbon follows. The foundation level of the Federation Tower is in a complex alternating sequence of variably intensively fissured, cavernous and porous limestone and variably hard, more or less watertight clay/marl. The thickness of the layers varies between 3 m and 10 m. The project area is located in a territory where potentially dangerous karst-suffusion processes occur.

Figure 21.
Federation Tower in Moscow City, Russia.

In the project, area exists several groundwater horizons carrying confined water which are not or just moderately corresponding with each other due to the sealing effect of the clay/marl. The pressure of the confined groundwater is up to 12 m. The groundwater mainly circulates in the fissured and karst-suffusion-affected limestone.

For the determination of the load-bearing behavior of deep foundation elements, two pile load tests have been carried out on the construction site. The test piles TP-15-1 and TP-15-2 had a diameter of 1.2 m and were instrumented with O-Cells. The pile segments in total were 6.9 m and 13.35 m long. The empty drill hole was filled with sand. The piles are completely positioned in the limestone (**Figure 22**).

The test piles had two segments with an O-cell in between. The displacements of the segments were measured with displacement transducers.

The maximum load of pile load test TP-15-1 was about 33 MN with an unloading-phase at 15 MN back to zero and a reloading-phase as shown in **Figure 23**. The upper

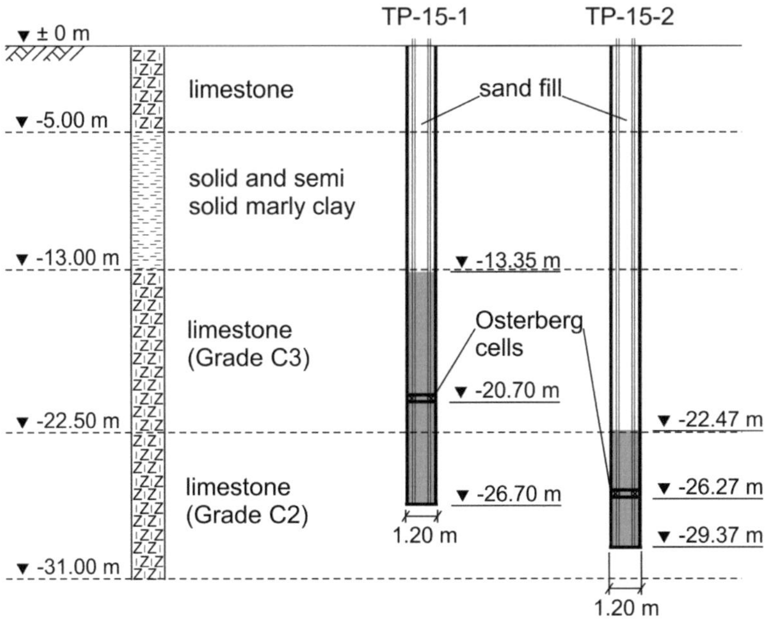

Figure 22.
Test piles TP-15-1 and TP-15-2 with O-Cells.

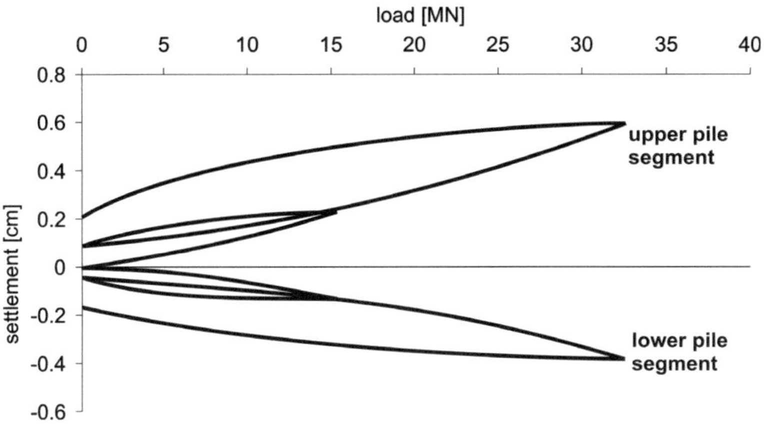

Figure 23.
Load-displacement diagram of test pile TP-15-1.

pile segment has a final displacement of 0.6 cm and the lower pile segment has a final displacement of 0.4 cm. No failure was seen and the empirically defined limit in [4, 6] of the settlement s = 0.1, D = 12 cm was not reached. The results of the pile load test TP-15-1 gives skin friction of q_s = 1140 kN/m^2 and base resistance of q_b = 5380 kN/m^2. Both values are not ultimate ones because failure criteria was not reached.

The maximum load of pile load test TP-15-2 was about 33 MN with three unloading-phases back to zero as shown in **Figure 24**. The upper pile segment has a final displacement of 4.3 cm and the lower pile segment has a final displacement of 2.2 cm. Again, no failure was seen and the empirically defined limit of the settlements of s = 0.1, D = 12 cm was not reached. The results of the pile load test TP-15-2 gives

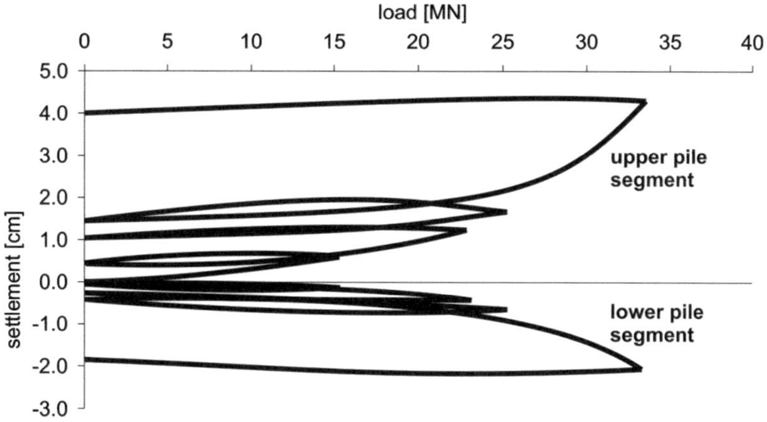

Figure 24.
Load-displacement diagram of test pile TP-15-2.

skin friction of q_s = 2310 kN/m^2 and base resistance of q_b = 5630 kN/m^2. Both values are not ultimate ones because failure criteria was not reached.

5. Summary and conclusions

The Combined Pile-Raft Foundation (CPRF) is a hybrid foundation system that combines the bearing capacity of a foundation raft and of piles or barrettes. The experiences made during the construction of several high-rise buildings show, that compared to a raft foundation a CPRF reduces the settlements by more than 50%. In addition, a CPRF reduces the necessary construction material including concrete and steel. This leads to a significant reduction of CO_2 emissions. To sum up the positive effects of a CPRF are:

- Increase of the overall stability of a raft foundation due to the reduction of the settlements, differential settlements, and tilts.

- Reduction of the inner forces and bending moments of the foundation raft using an optimized number and configuration of the piles.

- At foundation systems with an eccentricity the foundation resistance can be concentrated under the total building load; normally joints between the building elements are not necessary.

- Reduction of the uplift in the area of the excavation, because the relaxation of the soil is constrained.

- Cost optimization of the whole foundation system regarding the material used, time spent for construction, and CO_2 emitted.

Author details

Rolf Katzenbach[1] and Steffen Leppla[2*]

1 Ingenieursozietaet Professor Dr.-Ing. Katzenbach GmbH, Frankfurt am Main, Germany

2 Frankfurt University of Applied Sciences, Germany

*Address all correspondence to: steffen.leppla@fb1.fra-uas.de

References

[1] Aldrian W, Bantle A. Ways to reduce CO_2 in sprayed concrete for tunnel construction. Tunnel. 2021;**5**:36-46

[2] Katzenbach R, Leppla S, Choudhury D. Foundation Systems for High-rise Structures. New York, USA: CRC Press; 2016

[3] Katzenbach R. Optimized design of high-rise building foundations in settlement-sensitive soils. In: Proceedings of the International Geotechnical Conference of Soil-Structure-Interaction; 26–28 May 2005; St. Petersburg, Russia. Rotterdam, Netherlands: IOS Press; 2005. pp. 39-46

[4] Deutsche Gesellschaft für Geotechnik e.V. Recommendations on Piling. Berlin, Germany: Ernst & Sohn; 2018

[5] Kempfert HG, Moormann C. Pfahlgruendungen. In: Grundbau-Taschenbuch. Vol. 3. Berlin, Germany: Ernst & Sohn; 2018. pp. 79-323

[6] Deutsches Institut für Normung e.V. DIN EN 1997-1 Eurocode 7: Geotechnical design – Part 1: General rules. Berlin, Germany: Beuth; 2014

[7] International Society for Soil Mechanics and Geotechnical Engineering. ISSMGE Combined Pile-Raft Foundation Guideline. Darmstadt, Germany: Technical University Darmstadt; 2013

[8] Cooke RW. Piled raft foundations on stiff clays: A contribution to design philosophy. Géotechnique. 1986;**36**: 169-203

[9] Poulos HG. Pile behavior: Theory and application. Géotechnique. 1989;**39**: 365-415

[10] Randolph MF. Design methods for pile groups and piled rafts. In: Proceedings of the 13th International Conference on Soil Mechanics and Foundation Engineering; 5–10 January 1994; New Delhi, India. Vol. 5. London, United Kingdom: Taylor & Francis; 1994. pp. 61-82

[11] Poulos HG, Small JC, Ta LD, Simha J, Chen L. Comparison of some methods for analysis of piled rafts. In: Proceedings of the 14th International Conference on Soil Mechanics and Geotechnical Engineering; 6–12 September 1997; Hamburg, Germany. Vol. 2. New York, USA: CRC Press; 1997. pp. 1119-1124

[12] Katzenbach R, Reul O. Design and performance of piled rafts. In: 14th International Conference on Soil Mechanics and Geotechnical Engineering; 6–12 September 1997; Hamburg, Germany. Vol. 4. New York, USA: CRC Press; 1997. pp. 2253-2256

[13] Horikoshi K, Randolph MF. A contribution to optimal design of piled rafts. Géotechnique. 1998;**48**:301-317

[14] Russo G, Viggiani C. Factors controlling soil-structure interaction for piled rafts. Darmstadt Geotechnics. 1998;**2**:297-321

[15] Katzenbach R, Leppla S, Ramm H, Seip M, Kuttig H. Design and construction of deep foundation systems and retaining structures in uran areas in difficult soil and ground water conditions. In: Proceeding of the 11th International Conference on Modern Building Materials, Structures and Techniques; 16–17 May 2013; Vilnius, Lithuania. Kidlington, United Kingdom: Elsevier; 2013. pp. 540-548

[16] Turek J, Katzenbach R. New exhibition hall 3 in Frankfurt: Case

history of a Combined Pile-Raft Foundation subjected to horizontal load. In: Proceedings of the 5th International Conference on Case Histories in Geotechnical Engineering; 13–17 April 2004. New York, USA: CRC Press; 2004. pp. 1-5

[17] Katzenbach R, Leppla S, Ramm H. Combined Pile-Raft Foundation: Theory and practice. In: Design and Analysis of Pile Foundations. Boudouaou, Algeria: Dar Khettab Press; 2014. pp. 262-291

[18] Katzenbach R, Leppla S, Vogler M, Dunaevskiy R, Kuttig H. State of practice fort he cost-optimized foundation of high-rise buildings. In: Proceedings of the International Conference Geotechnical Challenges in Megacities; 7–10 June 2010; Moscow, Russia. Vol. 1. Melbourne, Australia: GRF; 2010. pp. 120-129

Chapter 5

Support Strength Criteria and Intelligent Design of Underground Powerhouses

Jianhai Zhang, Tianzhi Yao, Li Qian, Zuguo Mo, Yunpeng Gao, Fujun Xue, Chenggang Liao and Zhong Zhou

Abstract

The proper design of underground powerhouse support is the key engineering technique to guarantee the safe construction and operation of underground works. By regression analysis of the surrounding rock support parameters of 29 underground powerhouses with a span of 18.0–34.0 m, the empirical formula of the relationship between the support strength of anchor bar, strength-stress ratio, and plant span and the relationship among the support strength of the anchor cable, strength-stress ratio, and plant span are proposed. Furthermore, an intelligent design model for the anchor support of the underground powerhouse was trained by a BP (back propagation) neural network. Research shows that the support strength index of the anchor bolt and the anchor cable of these 29 plants are all distributed around 1.0. Therefore, a support strength index of 0.8–1.2 can be used as a reference for practical engineering support design. Finally, the reliability of the intelligent design model for the anchor support of the underground powerhouse was verified by comparison with actual engineering and support strength index. This shows that the intelligent design model can provide a reference for engineering design and construction.

Keywords: underground powerhouse, support strength criteria, strength-stress ratio, BP neural network, intelligent design

1. Introduction

The underground plant of a hydropower station is a large, complex underground building structure, and its stability is affected by factors such as geological structure, carven span, in situ stress, and support strength [1]. As underground plants are located in different stress environments, the lithology and strength of the surrounding rock are different, and the strength of the support to maintain the stability of the surrounding rock varies. Insufficient support strength can lead to local instability, collapse, excessive deformation of the surrounding rock, or even integral damage, while too much support strength can lead to unnecessary waste. Due to the complexity of the surrounding rock, scholars are still unable to fully grasp the deformation characteristics and reinforcement mechanism of the surrounding rock under complex stress conditions, which makes the theory and specification of surrounding rock reinforcement immature, and the support of underground plants still mainly relies on experience for design and construction. At present, the support design of underground plants is commonly based on the engineering analogy method, and there is insufficient knowledge of the deformation characteristics of the surrounding rock in high in situ stress areas. Because of the lack of relevant design experience, it is not sufficient to fully guide the design of the cavern support. The empirical method sometimes leads to safety problems due to inadequate design support strength or waste of resources due to over-support.

On the other hand, many successful examples of underground plants provide valuable data for the design of rock support. Through these data, the reinforcement measures and strength of the surrounding rock can be summarized, and the inherent laws of rock support and a new support design method can be proposed. For underground plants, the commonly used method is the system anchor and anchor cable support method, which can give good play to the strength and its own bearing capacity of the surrounding rock [2]. Through studying research papers and design data, a systematic summarization of sidewall support schemes for 29 underground plants with a span range of 18.0–34.0 m and a strength-stress ratio range of 2.0–14.55 was carried out, and the regression fitting relationships between the strength of the system anchor bolts and cables and the strength-stress ratio of the surrounding rock and the plant span were proposed. Based on the regression fitting relationship, an underground plant support strength index was defined, which can quantitatively evaluate whether the surrounding rock support is reasonable.

Neural network theory is recognized as a method for solving nonlinear problems, and it has been applied in rock mechanics parameter identification and stress analysis, parameter prediction, rock stability, rock deformation prediction, and rock engineering inverse analysis [3, 4]. One of the most popularly used neural network models is BP (back propagation) neural networks, which are multilayer feed-forward neural networks that are widely used in nonlinear modeling, function approximation, logic classification, etc. On this basis, an intelligent design model for the anchor support of the underground powerhouse is proposed based on a BP neural network. The model optimized the design of the system anchor diameter and spacing by inputting the plant span and strength-stress ratio. The different degrees of influence of the plant span and strength-stress ratio on the system anchor support scheme were analyzed according to the weights between the neurons.

1.1 Underground plant and rock surrounds support

1.1.1 Underground plant

According to incomplete statistics, more than 600 underground hydropower plants have been built worldwide, including more than 200 in Norway, which is the largest number of underground hydropower plants, and there are two underground power plants with an installed capacity of more than 1000 MW. As of 2015, the top 10 underground hydropower plants of installed capacity that have been built in the world are shown in **Table 1**.

The underground plant caverns are generally located in the hills downstream of the dam, mainly consisting of the diversion cavity, underground plant, traffic cavity, transformer room, surge chamber, and tailwater cavity, as shown in **Figure 1**. When designing the location of the plant, the longitudinal axis of the plant should have a small angle to the direction of the maximum principal stress of the initial ground stress and a large angle to the main structural surface, which is conducive to the stability of the cavern envelope.

Number	Name	Country	Installed capacity/ MW	Size of underground plant (L*W*H)/m	Completion date
1	Xiluodu	China	13,860	439.7*31.9*75.6	2014
2	Longtan	China	6300	388.5*30.3*74.5	2009
3	Nuozhadu	China	5850	418.0*29.0*79.6	2014
4	La Grande II	Canada	5280	490.0*26.3*47.2	1980
5	Churchill Falls	Canada	5225	300.0*24.5*45.5	1971
6	Jinping II	China	4800	352.4*28.3*72.2	2014
7	Sanxia	China	4200	311.3*32.6*87.3	2009
8	Xiaowan	China	4200	298.4*30.6*79.3	2012
9	Laxiwa	China	4200	311.7*30.0*73.8	2011
10	Jinping I	China	3600	277.0*28.9*68.8	2014

Table 1.
Top 10 installed capacity underground hydropower plants in the world.

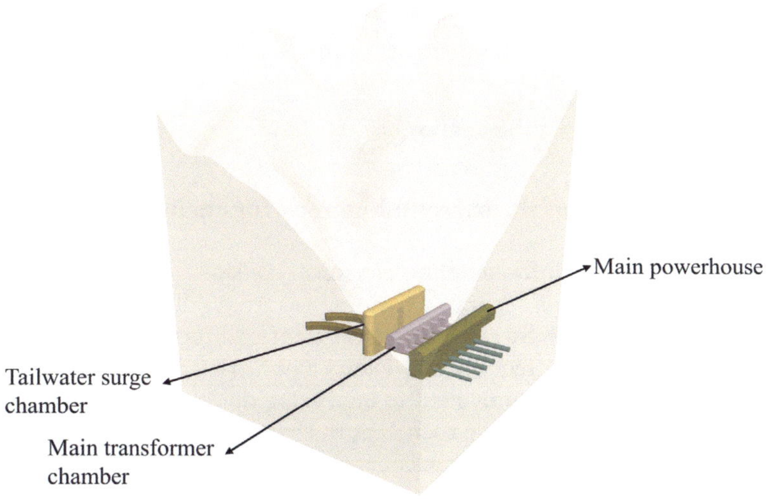

Figure 1.
Composition of the underground plant of hydropower station.

2. The effect of anchoring measures on the parameters of surrounding rock

2.1 Effect of anchor on the parameters of the surrounding rock

At present, numerical calculations generally take the anchor bolt (anchor cable) as the rod element, and the effect of the anchor bolt is reflected by the stiffness of the anchor, which is very small compared to the stiffness of the surrounding rock. Many calculations have shown that this method of simulation does not fully reflect the support effect of the anchor bolts. In fact, the main role of the anchor is to participate in the deformation process of the surrounding rock. The elastic recovery deformation of the anchor has a reverse locking force, which can create an anchoring effect on the surrounding rock. In other words, the deformation and strength parameters of the anchored rock mass can be increased and have been confirmed by laboratory and field tests [5].

For the strength of the surrounding rock after anchoring, the parameters for the shear strength of the surrounding rock after the anchor is applied can be calculated as:

$$C_1 = C_0 + \eta \frac{\tau_s S}{ab} \qquad (1)$$
$$\varphi_1 = \varphi_0$$

where C_0 and φ_0 are the cohesion and angle of internal friction of the surrounding rock before anchoring, respectively; τ_s is the shear strength of the anchor bolt; S is the cross-sectional area of the anchor bolt; a and b are the spacing and row spacing of the anchor arrangement, respectively; and η is the anchor group effect factor, which is dimensionless and is related to factors such as the anchor diameter, generally taken as $\eta = 2.0 \sim 5.0$. Eq. (1) shows that the improvement of the parameters of the surrounding rock by the anchor is mainly manifested by an increase in cohesion, and the increase in cohesion after the application of the anchor is as follows:

$$\Delta C_b = \eta \frac{\tau_s S}{ab} = \eta \tau_s \frac{\pi d^2}{4ab} \qquad (2)$$

where d is the diameter of the anchor.

2.2 Incremental cohesion of the surrounding rock for anchor cable reinforcement

The traditional anchor reinforcement mechanism considers the reinforcing effect of anchor cables as (1) keeping separated rock masses from falling off and (2) increasing the overall strength by rebounding the damaged rock masses. The anchor cable not only has the above effect but also exerts a positive pressure on the rock in the direction of the anchor. This is equivalent to increasing the lateral pressure of the surrounding rock, which changes the rock near the excavation face from a one-dimensional stress state to a three-dimensional stress state and increases the strength of the surrounding rock.

As shown in **Figure 2**, the state of the point on the free surface is one-dimensional pressure, that is, $\sigma_1 > 0$, and $\sigma_3 = 0$, corresponding to Mohr's circle **O**. The increase in wall pressure and the decrease in the radius of Mohr's circle after the application of the

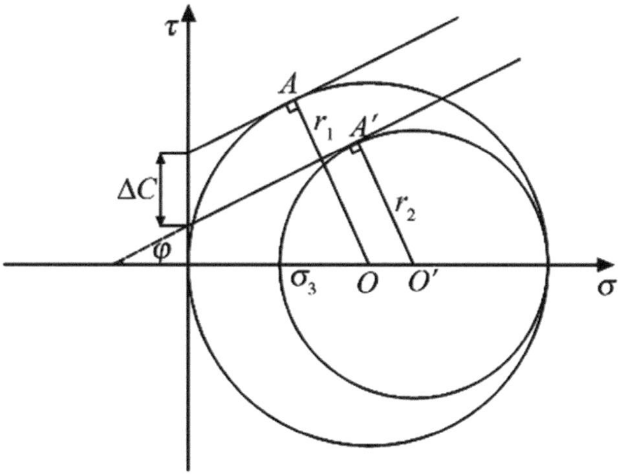

Figure 2.
Reinforcement mechanism of anchor cable.

prestress leads to a decrease in the tangent point of Mohr's circle from A to A′, which corresponds to an intercept difference ΔC with the τ-axis of the shear stress and is taken as the incremental cohesion ΔC_p of the surrounding rock provided by the anchor cable.

Assuming that the coefficient of friction $f = \tan\varphi$ of the rock mass is constant before and after reinforcement, it can be deduced from **Figure 2** that the cohesion of the rock mass can be increased by applying a prestressing force N (kN) with spacing a(m) and row b(m):

$$\Delta C_p = \eta \frac{Nf}{2ab}\left(1 + \frac{1}{\sin\varphi}\right) \tag{3}$$

Similar to Eq. (1), the anchor group effect factor $\eta = 2.0 \sim 5.0$, where φ is the internal friction angle of the surrounding rock before anchoring.

2.3 Comparison of the stability of the surrounding rock with and without support

Systematic support has a very significant effect on maintaining the stability of the surrounding rock during excavation. Taking the Yebatan hydropower station as an example, the distribution characteristics of the large deformation zones in the surrounding rock with and without system support were compared based on the FLAC^{3D} calculation software. The deformation distribution characteristics of the main powerhouse, main transformer chamber, and tailwater surge chamber under unsupported and systematically supported are shown in **Figures 3–10**. The comparison shows that:

1. The maximum local deformation of the roof arch of the main powerhouse is reduced from 70 ~ 130 mm to 60 ~ 80 mm and the maximum local deformation of the side walls is reduced from 120 ~ 180 mm to 100 ~ 150 mm under systematic support.

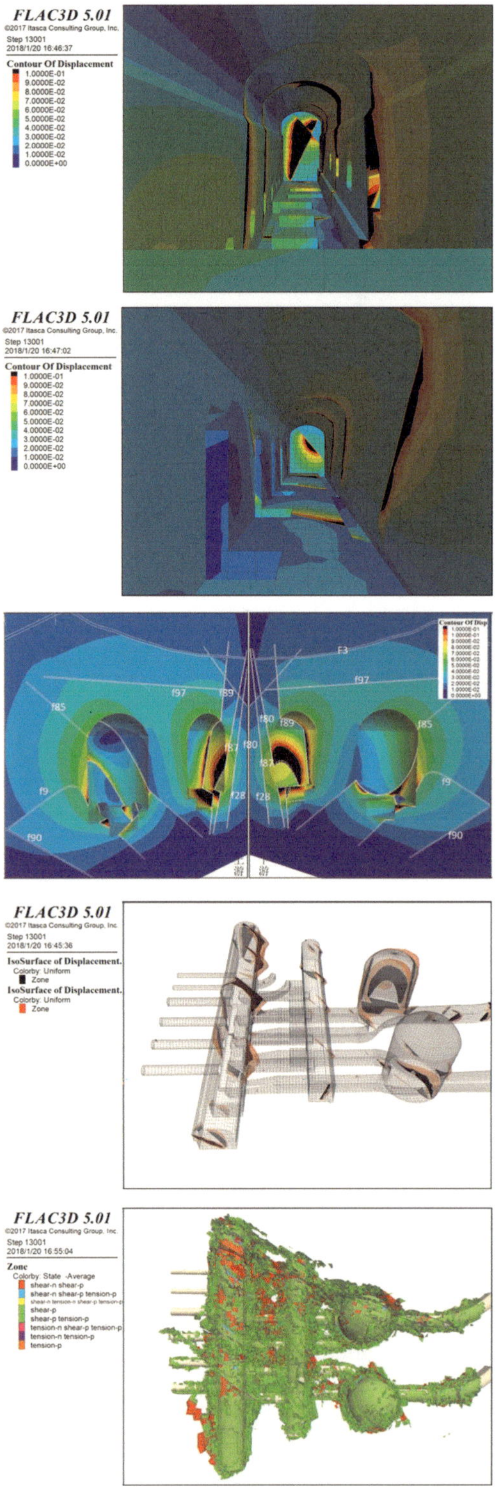

Figure 3.
Distribution characteristics of the deformation (black >100 mm) of the main powerhouse under unsupported conditions.

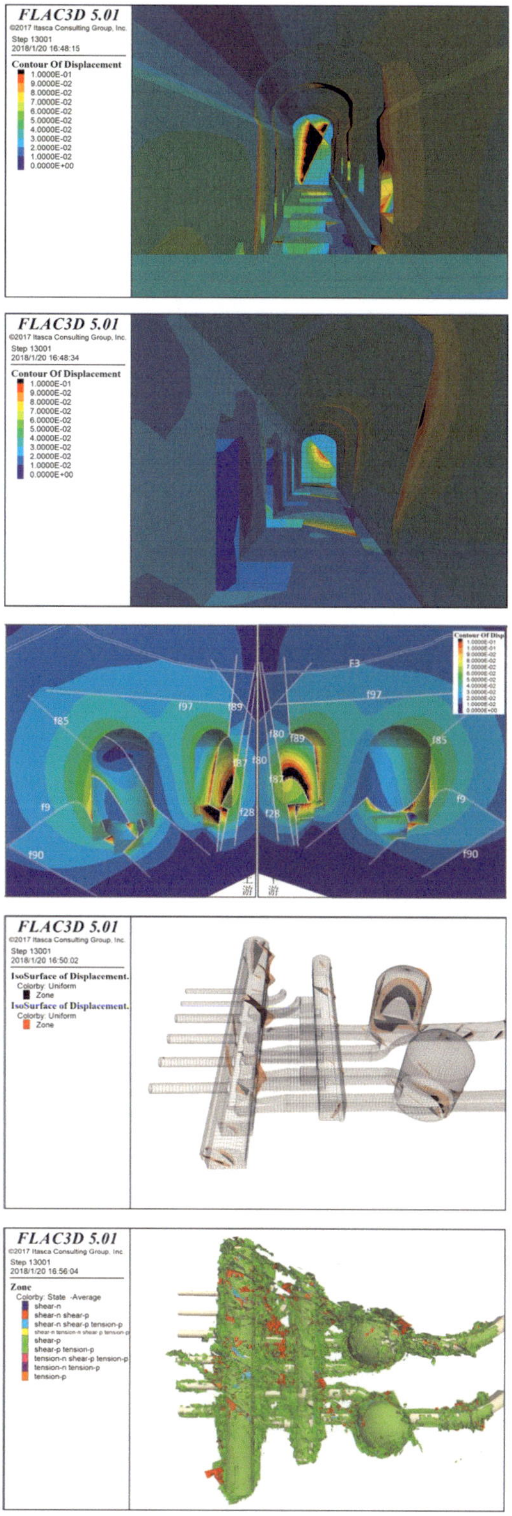

Figure 4.
Distribution characteristics of the deformation (black >100 mm) under systematic support conditions.

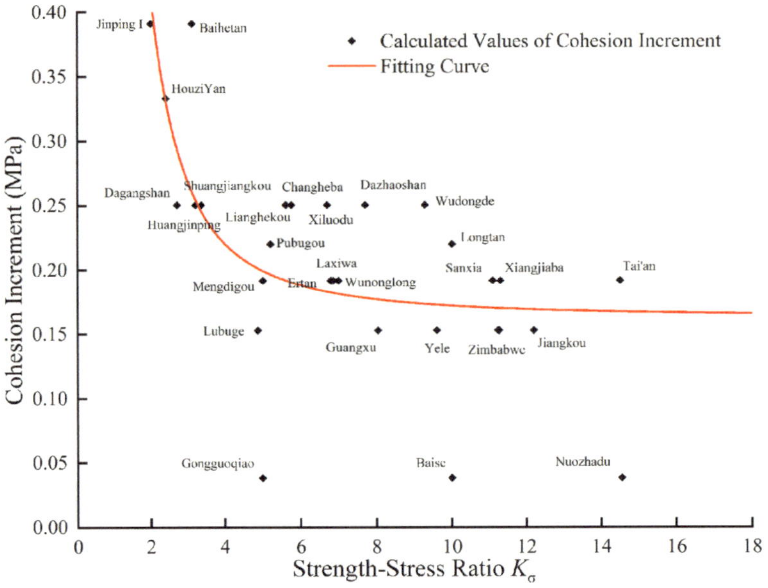

Figure 5.
Relationship between cohesion increment of the anchor bolt and strength stress ratio.

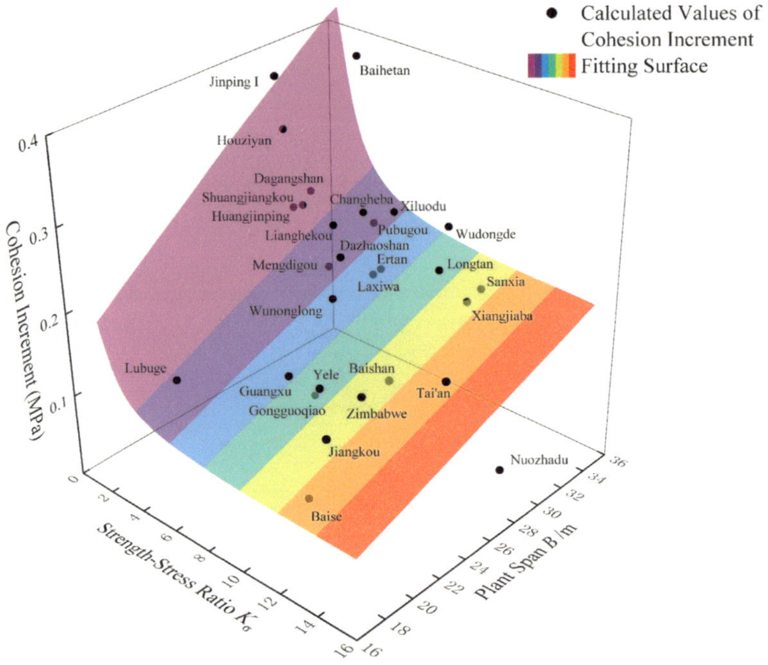

Figure 6.
Relationship between cohesion increment reinforced by anchor, strength-stress ratio, and plant span.

2. The maximum local deformation of the roof arch of the main transformer chamber is reduced from 65 ~ 85 mm to 50 ~ 60 mm; the maximum local deformation of the side walls is reduced from 90 ~ 135 mm to 70 ~ 110 under the systematic support.

DOI: http://dx.doi.org/10.5772/intechopen.102791

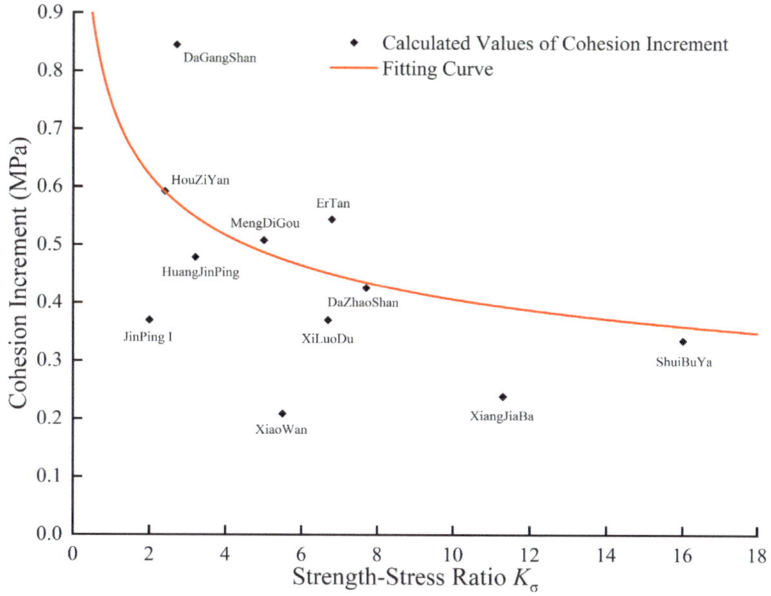

Figure 7.
Relationship between cohesion increment reinforces by anchor cables and strength-stress ratio.

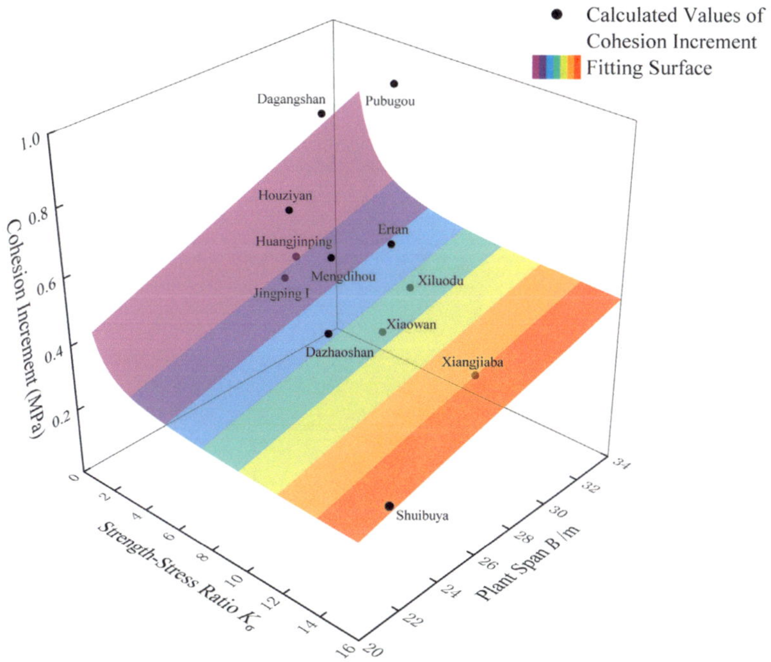

Figure 8.
Relationships among cohesion increment reinforce by anchor cable, strength stress ratio, and plant span.

3. The maximum local deformation of the roof arch of the tailwater surge is reduced from 80 ~ 110 mm to 70 ~ 105 mm, the maximum local deformation of the side walls is reduced from 100 ~ 170 mm to 100 ~ 130 mm. under the system support.

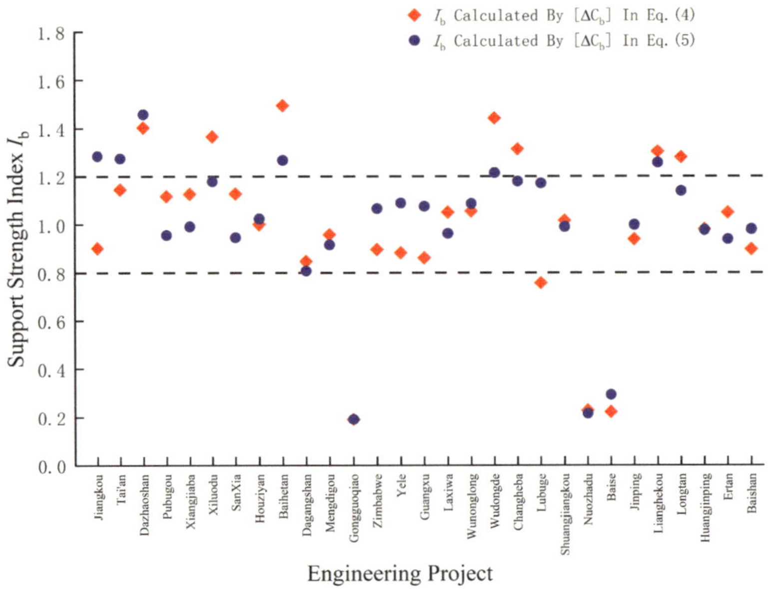

Figure 9.
Comparison of anchor bolt support index calculated by different fitting formulas.

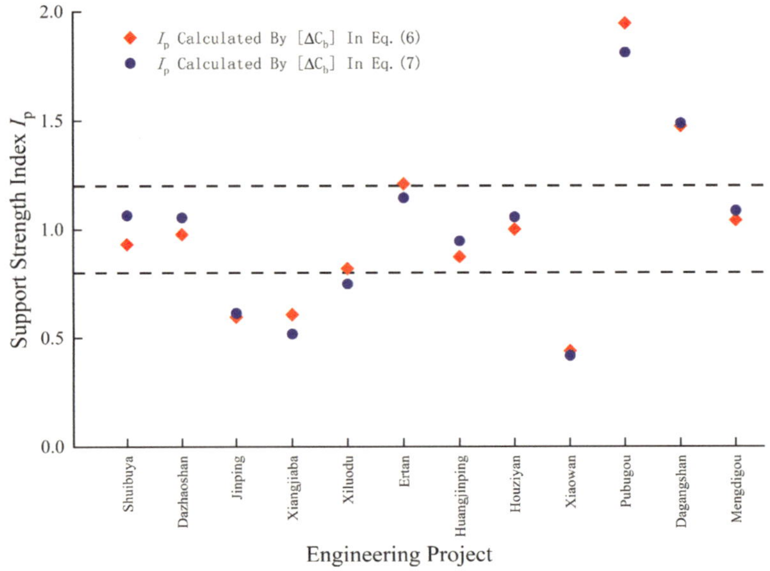

Figure 10.
Comparison of anchor cable support index calculated by different fitting formulas.

4. Under system support, the volume of the cavern group surrounding rock deformation greater than 100 mm is reduced from 21.6 to 9.7 thousand cubic meters.

In general, the deformation distribution characteristics of the surrounding rock under system support are similar to those under unsupported, but the extent and magnitude of large deformations at fault-affected areas are substantially reduced under system support.

3. Relationship among the plant span, strength-stress ratio, and incremental cohesion of the surrounding rock

3.1 Incremental cohesion of the surrounding rock after anchor reinforcement

As shown in **Table 2**, the physical and mechanical parameters, maximum principal stress, and anchor parameters of the underground plants of 29 hydropower stations, such as Jiangkou, Xiluodu, and Jinping I, are shown [6–36]. In **Table 2**, the maximum principal stress is taken as the maximum value near the main powerhouse. In accordance with the design principle of the "difference between arch and side wall", generally, the system anchors on the side wall of the main plant are used for statistical data. When the system anchor arrangement for the upstream and downstream side walls varies, the average value is taken as the statistical data. The incremental cohesion of the surrounding rock is calculated by Eq. (2), where η = 3.5, τ_s = 200 MPa, and the strength-stress ratio in **Table 2** is a dimensionless constant.

3.1.1 Relationship between the cohesion increment and strength-stress ratio of surrounding rock

The strength-stress ratio and cohesion increment of the 29 underground plants in **Table 2** are plotted as 29 data points in **Figure 5**. These data are fitted by a least-squares curve to obtain Eq. (4).

$$[\Delta C_b] = 0.45\left(2K_\sigma^{-2} + K_\sigma^{-4}\right) + 0.163 \tag{4}$$

As shown in **Figure 5**, most of the 29 data points are distributed near the fitted curve, forming a data band around a certain distance above and below the curve, and the cohesion increment of the surrounding rock increases with a decrease in the strength-stress ratio. The trend of the curve in **Figure 5** shows that when the strength-stress ratio $K_\sigma > 6.0$, the support strength reflected by the cohesion increment of the surrounding rock gradually tends to be constant. However, when the strength-stress ratio ranges from $3.0 \le K_\sigma < 6.0$, the curve gradually rises, indicating that the required support strength of the surrounding rock increases significantly as the strength-stress ratio decreases. When the strength-stress ratio $K_\sigma < 3$, the underground plant surrounding rock is in a high-very high-stress state, requiring an even higher support strength, and the cohesion increment ΔC_b and strength-stress ratio show -2 times nonlinearity. Eq. (4) shows that the smaller the strength of the surrounding rock and the higher the in situ stress of the underground plant are, the greater the support strength required, but the growth rate shows a nonlinear relationship with the strength-stress ratio.

3.1.2 Relationship between the cohesion increment and the strength-stress ratio and plant span

According to the 29 underground plants in **Table 2**, the cohesion increment ΔC_b of the surrounding rock is fitted to the plant span B and the strength-stress ratio K_σ by least squares surface fitting, and the equation is as follows:

$$[\Delta C_b] = 0.01481\left(2K_\sigma^{-2} + 0.405\right)B \tag{5}$$

Engineering	Cavern span/m	Uniaxial compressive strength/ MPa	Maximum principal stress/ MPa	Strength-stress ratio K_σ	Anchor diameter D/mm	Anchor spacing and row spacing a, b/m	Calculated values of cohesion increments ΔC_b/MPa
Jiangkou	19.2	90	7.40	12.2	25.0	1.5	0.153
Tai'an	24.5	160	11.0	14.5	28.0	1.5	0.192
Dazhaoshan	26.4	85	11.0	7.70	32.0	1.5	0.250
Pubugou	32.4	120	23.3	5.20	30.0	1.5	0.220
Xiangjiaba	31.0	100	8.90	11.3	28.0	1.5	0.192
Xiluodu	31.9	120	18.0	6.70	32.0	1.5	0.250
SanXia	32.5	130	11.7	11.1	28.0	1.5	0.192
Houziyan	29.2	80	33.5	2.40	32.0	1.3	0.333
Baihetan	34.0	95	31.0	3.10	32.0	1.2	0.391
Dagangshan	30.8	60	22.2	2.70	32.0	1.5	0.250
Mengdigou	29.1	85	17.0	5.00	28.0	1.5	0.192
Gongguoqiao	27.8	70	14.0	4.99	25.0	3.0	0.038
Zimbabwe	23.0	100	8.90	11.24	25.0	1.5	0.153
Yele	22.2	120	12.5	9.60	25.0	1.5	0.153
Guangxu	22.0	112.5	14.0	8.04	25.0	1.5	0.153
Laxiwa	30.0	157.0	22.87	6.86	28.0	1.5	0.192
Wunonglong	26.7	70.0	10.0	7.00	28.0	1.5	0.192
Wudongde	32.5	90.0	9.70	9.28	32.0	1.5	0.250
Changheba	30.8	138	24.0	5.75	32.0	1.5	0.250
Lubuge	18.0	82.8	17.0	4.87	25.0	1.5	0.153
Shuangjiangkou	29.3	97.3	29.0	3.36	32.0	1.5	0.250
Nuozhadu	29.0	120.3	8.27	14.55	25.0	3.0	0.038
Baise	20.7	60.0	6.00	10.00	25.0	3.0	0.038
Jinping I	29.2	70	35.7	2.00	32.0	1.2	0.391
Lianghekou	28.7	100	18.0	5.60	32.0	1.5	0.250
Longtan	30.7	130	13.0	10.0	30.0	1.5	0.220
Huangjinping	28.8	75	23.2	3.20	32.0	1.5	0.250
Ertan	30.7	200	29.5	6.80	28.0	1.5	0.192
Baishan	25.0	108.0	9.58	11.27	25.0	1.5	0.153

Table 2.
Relevant data of the powerhouse and calculation of the cohesion incrementally reinforced by the anchor bar.

In **Figure 6**, the cohesion increment provided by the anchor is approximately linearly related to plant span B, which increases with an increase in plant span. In addition, the cohesion increment ΔC_b and strength-stress ratio still show -2 times nonlinearity.

DOI: http://dx.doi.org/10.5772/intechopen.102791

3.2 Incremental cohesion of the surrounding rock for anchor cable reinforcement

The anchor cable support parameters for 12 large and medium-sized hydropower plants are shown in **Table 3**, and the cohesion increment in the surrounding rock is calculated by Eq. (3), where η = 3.5.

3.2.1 Relationship between the cohesion increment and strength-stress ratio of the surrounding rock

The 12 points are plotted in **Figure 7** based on the strength-stress ratio and cohesion increment ΔC_p of the surrounding rock for the 12 underground plants in **Table 3**. Eq. (6) can be obtained by fitting the least squares curve to the 12 points:

$$[\Delta C_{\mathrm{p}}] = 0.7445 K_{\sigma}^{-0.2627} \tag{6}$$

As shown in **Figure 7**, the incremental cohesion ΔC_p of the surrounding rock provided by the anchor cable decreases with increasing strength-stress ratio. When the strength-stress ratio $K_\sigma \geq 4.0$, the weakening rate of the support strength gradually decreased at a certain rate. When $K_\sigma < 4.0$, the increase rate of the support strength accelerates.

A comparison of the fitting curves of the anchor bolt in **Figure 5** and the anchor cable in **Figure 7** shows the following differences: (1) There is no obvious transition zone in the fitted curve of the anchor cable; (2) When $K_\sigma < 6.0$, the upward trend of the fitted curve of the anchor cable is less than that of the anchor bolt; and (3) when $K_\sigma \geq 6.0$, the fitted curve of the anchor cable does not converge to a constant as the anchor bolt fitting curve does but decreases at a certain rate. These differences indicate that anchor cables provide greater support strength than anchor bolts and that the rate of change in anchor cable support strength with strength-stress ratio is less than that of the anchor bolts at $K_\sigma < 4.5$.

3.2.2 Relationship among the cohesion increment, strength-stress ratio, and plant span

The relationship among cohesion increments ΔC_p, plant span B, and the strength-stress ratio K_σ of the 12 underground plants are plotted in **Figure 8**. The equation can be obtained by least-squares surface fitting:

$$[\Delta C_{\mathrm{P}}] = 0.00262\left(5.383 + 3K_{\sigma}^{-1} + 4K_{\sigma}^{-2}\right)B \tag{7}$$

As shown in **Figure 8**, the 12 data points are distributed approximately around the fitted curve surface, and the incremental cohesion of the surrounding rock increases with the plant span, which is consistent with engineering practice. The incremental cohesion of the surrounding rock is approximately linearly related to the plant span when the strength-stress ratio is greater than a certain value.

3.3 Support strength criteria

To better reflect the relative relationship between the actual support strength and the empirical formula, the dimensionless anchor support strength index I_b is defined as follows:

Engineering project	Plant span/m	Uniaxial compressive strength/MPa	Maximum principal Stress/MPa	Strength-stress ratio K_σ	Anchor cable internal force/kN	Anchor spacing a/m	Anchor row spacing b/m	Calculated values of cohesion increments ΔC_p/MPa
Shuibuya	21.5	90	5.62	16.0	1500	4.2	4.5	0.335
Dazhaoshan	26.4	85	11.00	7.7	2000	4.5	5.2	0.426
Jinping I	29.2	70	35.70	2.0	1750	4.5	4.5	0.370
Xiangjiaba	31.0	100	8.85	11.3	1500	5.0	6.0	0.239
Xiluodu	31.9	120	18.00	6.7	1750	4.5	4.5	0.370
Ertan	30.7	200	29.54	6.8	1500	3.0	2.0	0.544
Huangjinping	28.8	75	23.23	3.2	1750	4.0	4.0	0.479
Houziyan	29.2	80	33.45	2.4	2500	4.0	4.0	0.592
Xiaowan	31.5	140	25.40	5.5	1000	5.0	5.0	0.209
Pubugou	32.4	120	23.30	5.2	2000	3.0	3.0	0.939
Dagangshan	30.8	60	22.90	2.7	1800	4.5	4.5	0.845
Mengdigou	29.1	85	17.00	5.0	2000	4.5	4.5	0.508

Table 3.
Relevant data of each powerhouse and calculations of the cohesion increment reinforced by the anchor cable.

$$I_b = \frac{\Delta C_b}{[\Delta C_b]} \tag{8}$$

where the numerator is the calculated value of the design anchor support strength, which is calculated by Eq. (2); and the denominator is the support strength calculated by empirically fitting Eqs. (4) or (5).

Engineering	Calculated cohesion increment ΔC_b/MPa	Values calculated by Eq. (4) $[\Delta C_b]$/MPa	Values calculated by $[\Delta C_b]$ of Eq. (4) I_b	Values calculated by Eq. (5) $[\Delta C_b]$/MPa	Values calculated by $[\Delta C_b]$ of Eq. (5) I_b
Jiangkou	0.153	0.169	0.90	0.119	1.28
Tai'an	0.192	0.167	1.15	0.150	1.27
Dazhaoshan	0.250	0.178	1.40	0.172	1.46
Pubugou	0.220	0.197	1.12	0.230	0.96
Xiangjiaba	0.192	0.170	1.13	0.193	0.99
Xiluodu	0.250	0.183	1.37	0.212	1.18
SanXia	0.192	0.170	1.13	0.203	0.95
Houziyan	0.333	0.333	1.00	0.325	1.02
Baihetan	0.391	0.262	1.49	0.309	1.27
Dagangshan	0.250	0.295	0.85	0.310	0.81
Mengdigou	0.192	0.200	0.96	0.209	0.92
Gongguoqiao	0.038	0.200	0.19	0.200	0.19
Zimbabwe	0.153	0.170	0.90	0.143	1.07
Yele	0.153	0.173	0.88	0.140	1.09
Guangxu	0.153	0.177	0.86	0.142	1.08
Laxiwa	0.192	0.182	1.05	0.199	0.96
Wunonglong	0.192	0.182	1.06	0.176	1.09
Wudongde	0.250	0.174	1.44	0.206	1.21
Changheba	0.250	0.191	1.31	0.212	1.18
Lubuge	0.153	0.202	0.76	0.130	1.17
Shuangjiangkou	0.250	0.246	1.02	0.253	0.99
Nuozhadu	0.038	0.167	0.23	0.178	0.21
Baise	0.038	0.172	0.22	0.130	0.29
Jinping I	0.391	0.416	0.94	0.391	1.00
Lianghekou	0.250	0.192	1.30	0.199	1.26
Longtan	0.220	0.172	1.28	0.193	1.14
Huangjinping	0.250	0.255	0.98	0.256	0.98
Ertan	0.192	0.183	1.05	0.204	0.94
Baishan	0.153	0.170	0.90	0.156	0.98

Table 4.
Anchor bolt support index of the underground powerhouse calculated by Eqs. (4) and (5).

Similarly, the dimensionless anchor cable support strength index I_p can be defined as:

$$I_p = \frac{\Delta C_p}{[\Delta C_p]} \tag{9}$$

where the numerator is the calculated value of the design anchor cable support strength, which is calculated by Eq. (3); and the denominator is the support strength calculated by empirically fitting Eqs. (6) or (7).

The cohesion increment of the anchor cable are calculated by Eqs. (4) and (5) and the anchor cable support strength index calculated by Eq. (8) for each engineering are shown in **Table 4**. **Figure 9** shows the comparison between I_b calculated by $[\Delta C_b]$ of Eq. (4) and I_b calculated by $[\Delta C_b]$ of Eq. (5). It can be seen from **Figure 9** that (1) the anchor bolt support index I_b is mostly distributed at approximately 1.0, in which 68.96% of I_b calculated by Eq. (5) and 65.51% I_b calculated by Eq. (4) are between 0.8 ~ 1.2; and (2) I_b calculated by Eq. (5) is closer to 1 than I_b calculated by Eq. (4). This shows that Eq. (5), which considers both the plant span and the strength-stress ratio, can better reflect the support strength of the anchors.

The cohesion increment of the anchor cable calculated from Eqs. (6) and (7) for each engineering project and the anchor cable support index calculated using Eq. (9) for each project are shown in **Table 5**. The comparison between I_p calculated by $[\Delta C_p]$ of Eq. (6) and I_p calculated by $[\Delta C_p]$ of Eq. (7) is shown in **Figure 7**. It can be seen from **Figure 10** that (1) the anchor cable support index I_p is mostly distributed at approximately 1.0, in which 58.3% I_b is between 0.8 and 1.2; and (2) the support indices calculated by different fitting formulas for the same engineering are similar. This shows that the empirical formula can reflect the strength of the anchor cable support well, and the fitted results of empirical Eqs. (6) and (7) are similar.

Engineering	Calculated cohesion increment ΔC_p/MPa	Values calculated by Eq. (6) $[\Delta C_p]$/MPa	Values calculated by $[\Delta C_p]$ of Eq. (6) I_p	Values calculated by Eq. (7) $[\Delta C_p]$/MPa	Values calculated by $[\Delta C_p]$ of Eq. (7) I_p
Shuibuya	0.335	0.359	0.93	0.315	1.06
Dazhaoshan	0.426	0.435	0.98	0.404	1.05
Jinping I	0.370	0.621	0.60	0.603	0.61
Xiangjiaba	0.239	0.394	0.61	0.461	0.52
Xiluodu	0.370	0.452	0.82	0.495	0.75
Ertan	0.544	0.450	1.21	0.475	1.14
Huangjinping	0.479	0.548	0.87	0.506	0.95
Houziyan	0.592	0.592	1.00	0.561	1.06
Xiaowan	0.209	0.476	0.44	0.500	0.42
Pubugou	0.939	0.483	1.94	0.518	1.81
Dagangshan	0.845	0.574	1.47	0.568	1.49
Mengdigou	0.508	0.488	1.04	0.468	1.08

Table 5.
Anchor cable support strength index of the underground powerhouse calculated by Eqs. (6) and (7).

In summary, combined with the actual engineering and experience formula, the support strength index can be used as a reference basis and judging standard for the actual engineering support design:

$$\left.\begin{array}{ll} I_P \text{ or } I_b < 0.8, & \text{Low Support Strength} \\ I_P \text{ or } 0.8 < I_b < 1.2, & \text{Reasonble Support Strength} \\ I_P \text{ or } I_b > 1.2, & \text{High Support Strength} \end{array}\right\} \tag{10}$$

4. Intelligent design model for the anchor support of the underground powerhouse

4.1 Model design and training logic

4.1.1 Model design

As seen from Parts 2 and 3, the design of the anchor support for the underground powerhouse can be determined by the plant span and strength-stress ratio. There is a certain relationship among the plant span, strength-stress ratio, anchor diameter, anchor spacing, and row spacing. Their mapping relationship can be reflected by a back propagation (BP) neural network. The BP neural network is a multilayer feed-forward neural network that is widely used in nonlinear modeling, function approximation, logic classification, etc. Therefore, an intelligent design model for anchor support of the underground powerhouse was created, which can output diameter D, anchor spacing a, and row spacing b by inputting plant span B and strength-stress ratio K_σ. The model takes advantage of the logical classification of BP neural networks to find the mapping of plant spans and strength-stress ratio to different support schemes. By analyzing the scheme of anchor bolt support of completed underground powerhouses in **Table 1**, the anchor bolt support schemes are divided into six types, as shown in **Table 6**.

The model consists of three parts: an input layer, a hidden layer, and an output layer. The structure of the model is shown in **Figure 11**. The input layer contains the plant span and strength-stress ratio, and the output layer contains the anchor diameter, spacing, and row spacing of the anchor. The hidden layer is used to connect the input and output layers and to pass the weights of the neural network. The number of layers and nodes in the hidden layers affect the prediction results of the model. Theoretically, the greater the number of hidden layers is, the smaller the error of the prediction results, but too many hidden layers will lead to an overly complex network

Scheme number	Anchor diameter D/mm	Anchor spacing and row spacing a, b/m
1	32	1.2
2	32	1.5
3	30	1.5
4	28	1.5
5	25	1.5
6	25	3.0

Table 6.
Anchor support schemes.

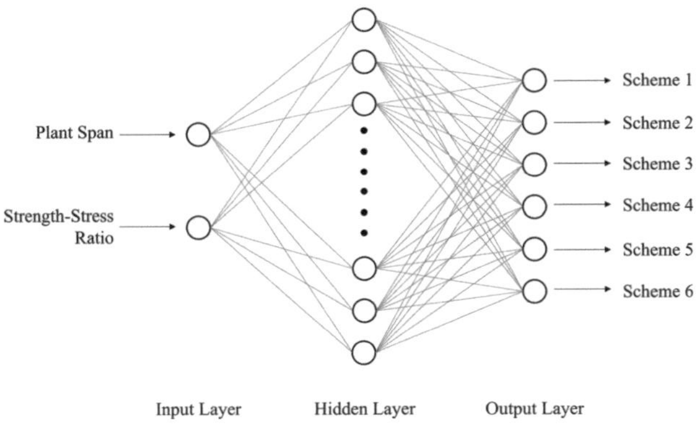

Figure 11.
The structure of BP neural network.

structure and slow computation speed. In this paper, the number of hidden layers is chosen as one layer with reference to a typical BP neural network structure. The number of nodes in the hidden layer is directly related to the number of input and output units, but there is still no perfect analytical formula. Too many nodes in the hidden layer will lead to a long learning time, while too few nodes in the hidden layer will have poor fault tolerance. According to previous experience [37], the number of nodes is designed with reference to Eq. (11).

$$m = \sqrt{n + l} + \alpha \tag{11}$$

where m is the number of nodes in the hidden layer, n is the number of nodes in the input layer, l is the number of nodes in the output layer, and α is a constant between 1 and 10. In this model, the value of α is 7, so the number of hidden layer nodes calculated according to Eq. (11) is 10.

4.1.2 Model training logic

1. Forward propagation

   The initial training of the model in Step 3 of Section 4.1.2 is achieved by forward propagation of the BP neural network. Suppose the sample set is $\mathbf{X}$, the second layer of the BP neural network (hidden layer) is $\mathbf{a}_2$, the third layer (output layer) is $\mathbf{a}_3$, $\mathbf{\Theta}^{(i)}$ is the weight from layer i to layer $(i + 1)$, the initial $\mathbf{\Theta}^{(i)}$ is set randomly, the model target output value is $\mathbf{y}$, and $\mathbf{h}$ is the actual output value of the model after training. Forward propagation can be expressed by the following equations:

$$\mathbf{a}_2 = sigmoid\left(\mathbf{\Theta}^{(1)} \times \mathbf{X}^{\mathrm{T}}\right) \tag{12}$$

$$\mathbf{a}_3 = sigmoid\left(\mathbf{\Theta}^{(2)} \times \mathbf{X}^{\mathrm{T}}\right) \tag{13}$$

$$\mathbf{h} = \mathbf{a}_3 \tag{14}$$

   where $\mathbf{X}$, $\mathbf{a}_2$, $\mathbf{a}_3$, $\mathbf{\Theta}^{(i)}$, and $\mathbf{h}$ are the matrix and *sigmoid* is the transfer function, as shown in Eq. (15).

$$sigmoid(x) = \frac{1}{1+e^{-x}} \tag{15}$$

2. Cost function

   Because the initial $\mathbf{\Theta}^{(i)}$ is set randomly, the actual output value **h** of the initial model has a large error with the target output value **y**. To evaluate the accuracy of the actual output value **h**, the cost function $J(\mathbf{\Theta})$ is introduced, and the formula is shown in Eq. (16). The smaller the value of $J(\mathbf{\Theta})$ is, the closer the actual output value **h** is to the target output value **y**, representing a better value for the weight $\mathbf{\Theta}$.

$$J(\mathbf{\Theta}) = -\frac{1}{m}\sum_{i=1}^{m}\sum_{k=1}^{K}\left[x_k^{(i)}\log\left(h_{\mathbf{\Theta}}\left(x^{(i)}\right)\right)_k + \left(1-y_k^{(i)}\log\left(1-\left(h_{\mathbf{\Theta}}\left(x^{(i)}\right)\right)_k\right)\right)\right] + \frac{\lambda}{2m}\sum_{l=1}^{L-1}\sum_{i=1}^{s_l}\sum_{j=1}^{s_{l+1}}\left(\mathbf{\Theta}_{ji}^{(l)}\right)^2 \tag{16}$$

   where $x(i)\ k$ is the k-th data in the i-th layer, m is the total number of layers in the BP neural network, K is the total number of data, λ is a constant, L is the total number of layers in the neural network, and s_l is the number of nodes in the l-th layer.

3. Back propagation

To continuously obtain a smaller cost function $J(\mathbf{\Theta})$, we continuously update the value of the weight $\mathbf{\Theta}$ by back propagation. The error transfer and weight update process are as follows:

$$\delta_k^{(3)} = \left(a_k^{(3)} - y_k\right) \tag{17}$$

$$\boldsymbol{\delta}^{(2)} = \left(\mathbf{\Theta}^{(2)}\right)^T \boldsymbol{\delta}^{(3)} \times \mathbf{g}\prime^{\left(z^{(2)}\right)} \tag{18}$$

$$\mathbf{g}\prime^{\left(z^{(2)}\right)} = \mathbf{a}^{(2)} \times \left(1 - \mathbf{a}^{(2)}\right) \tag{19}$$

where $\delta(l)\ j$ is the error of the j-th node in the l-th layer and $a(i)\ k$ is the k-th data in the i-th layer.

The errors are stored in $\mathbf{\Delta}^{(l)}$.

$$\mathbf{\Delta}^{(l)} = \mathbf{\Delta}^{(l)} + \boldsymbol{\delta}^{(l+1)}\left(\mathbf{a}^{(l)}\right)^T \tag{20}$$

where $\mathbf{\Delta}^{(l)}$ is the set of each node in l-th layer. The weight is updated with $\mathbf{\Delta}^{(l)}$ and calculated by Eq. (20). The intelligent design model is obtained after the optimal weights are calculated.

4.1.3 Model training

The first 23 data points in **Table 1** were used as training samples to train the intelligent design model for anchor support of the underground powerhouse. The training process of the model is shown in **Figure 12**.

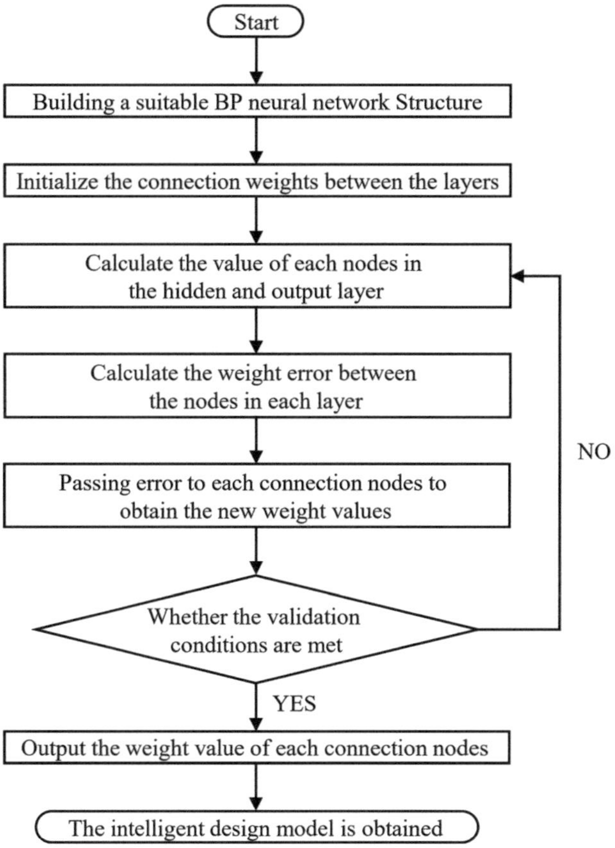

Figure 12.
The training process of BP neural network.

Step 1: A neural network structure applicable to the intelligent design model is established, as shown in **Figure 11**.
Step 2: The weight values of the neural network are initialized.
Step 3: The training set data are input. The initial training model is obtained by forward propagation of the neural network, and the values of the nodes in the neural network are obtained.
Step 4: The output layer errors of the initial training model are calculated and passed to the hidden layer to update the weights between the output layer and hidden layer. Similarly, the errors between the hidden layer and input layer are calculated, and the weights are updated.

Engineering	Plant span B/m	Uniaxial compressive strength R_c/MPa	Maximum principal stress/MPa	Strength-stress ratio K_σ	Anchor diameter D/mm	Anchor spacing and row spacing a, b/m
Jinping I	29.2	70	35.7	2.00	32.0	1.2
Lianghekou	28.7	100	18.0	5.60	32.0	1.5
Longtan	30.7	130	13.0	10.0	30.0	1.5

Table 7.
Validation data.

Engineering	Plant span B/m	Uniaxial compressive strength R_c/MPa	Maximum principal stress/MPa	Strength-stress ratio K_σ	Anchor diameter D/mm	Anchor spacing and row spacing a, b/m
Huangjinping	28.8	75	23.2	3.20	32.0	1.5
Ertan	30.7	200	29.5	6.80	28.0	1.5
Baishan	25.0	108.0	9.58	11.27	25.0	1.5

Table 8.
Testing data.

Step 5: After the new weights have been calculated, the new model is validated by the validation set, and if it does not meet the requirements of the validation set, then step 4 is repeated. If it does, then the final intelligent support model is obtained.

The validation set for the intelligent design model is shown in **Table 7**. The test set for the intelligent design model is shown in **Table 8**.

4.2 Model implementation

Based on the process in Section 4.1 and the data in **Table 1**, the intelligent design model was trained. Now, if the plant span and strength-stress ratio are input into the model, then the anchor diameter, spacing, and row spacing can be output. The interaction of the models can be implemented by MATLAB 2019b.

Taking the underground plant of Huangjinping Hydropower Station as an example, by inputting the plant span B and the strength-stress ratio K_σ, the model will automatically output the anchor support scheme, and the results are shown in **Figure 13**.

4.3 Model test and discussion

The test data are input in **Table 8** into the model, in turn, to obtain the support design scheme for the test set, as shown in **Table 9**.

As seen in **Table 9**, the scheme proposed by the system anchor design model for the Baishan Hydropower Station is consistent with the scheme used in actual engineering. The proposed scheme for the Ertan Hydropower Station is anchor diameter Φ30, and the anchor spacing and row spacing are @1.5 × 1.5. The proposed scheme for the Huangjinping Hydropower Station is anchor diameter Φ32, and the anchor spacing and row spacing are @1.2 × 1.2. The proposed schemes are safer and more reliable than the scheme used in actual engineering.

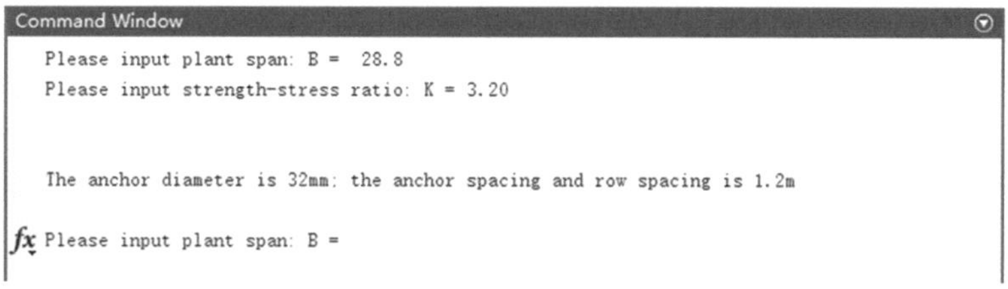

Figure 13.
Intelligent model run testing.

Engineering	Plant span B/m	Strength-stress ratio K_σ	Actual scheme of engineering	Scheme proposed by intelligent design model
Huangjinping	28.8	3.20	Anchor Diameter Φ32, Spacing and Row Spacing @1.5×1.5	Anchor Diameter Φ32, Spacing and Row Spacing @1.2×1.2
Ertan	30.7	6.80	Anchor Diameter Φ28, Spacing and Row Spacing @1.5×1.5	Anchor Diameter Φ30, Spacing and Row Spacing @1.5×1.5
Baishan	25.0	11.27	Anchor Diameter Φ25, Spacing and Row Spacing @1.5×1.5	Anchor Diameter Φ25, Spacing and Row Spacing @1.5×1.5

Table 9.
Comparison between the actual scheme and the proposed scheme of the intelligent design model.

The schemes suggested by the intelligent design model are evaluated using the concept of support strength criteria presented in Section 3.3. The comparison of the proposed schemes from the intelligent design model and the actual support schemes are shown in **Figure 14**.

In high in situ stress areas, the engineering analogy method has fewer projects to provide a reliable reference for support schemes with different plant spans and strength-stress ratios. Therefore, in actual engineering, there may be a situation where the design support strength is low, such as Dagangshan Hydropower Station in Figure 15, which may cause dangerous situations during construction. In low and medium in situ stress areas, the same low strength of system anchor support was observed in Mengdigou, Baise, and Nuozhadu hydropower stations designed by the traditional method. This indicates that the engineering analogy method has difficulty

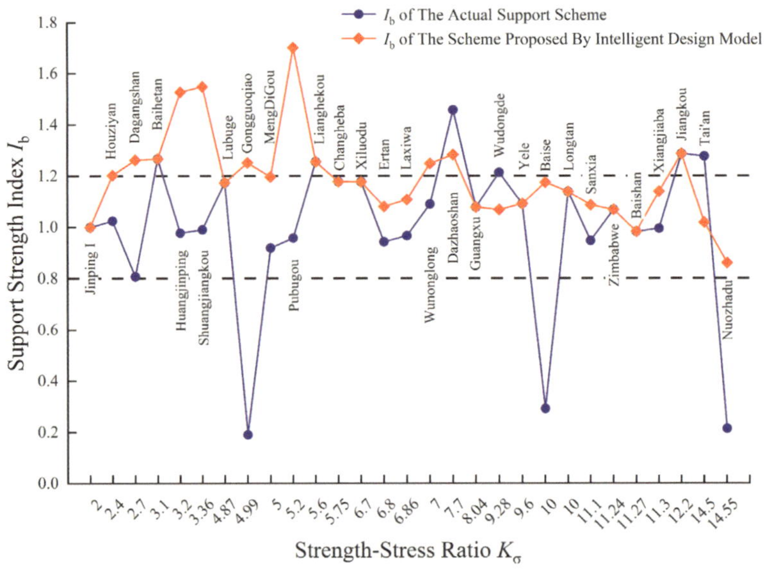

Figure 14.
Comparison of the support strength index of the actual scheme and the scheme proposed by the intelligent design model.

selecting an appropriate system anchor support scheme in situations such as high in situ stress and uncommon plant spans and that the design reliability is low. However, the support strengths suggested by the intelligent design model are generally better than those used in the actual project, and generally, the support index is above 1.0. In relatively complex areas of high in situ stress, such as the Houziyan and Dagangshan hydropower stations, intelligent design models provide safer design schemes than actual engineering. This shows that the intelligent design model can provide a more reliable and economical support scheme than the traditional engineering analogy method and can be used as a reference for the design of system anchors for underground plants in practical engineering.

5. Influencing factors of support scheme design

5.1 Evaluation methods for neural network weights

The choice of a system anchor support scheme in an underground plant is influenced by several factors, but it is not yet clear which is the main factor. Based on the intelligent design model proposed above, the different degrees of influence of the plant span and strength-stress ratio on the selection of the system anchor support scheme can be further explored by analyzing the weights between the neurons. The relationship can be described with the help of the following indicators [38].

1. Correlation significance coefficient r_{ij}:

$$r_{ij} = \sum_{k=1}^{p} \frac{\omega_{ki}(1-e^{-x})}{1+e^{-x}} \qquad x = \omega_{jk} \tag{21}$$

where i is the input unit of the neural network, $i = 1, \ldots \ldots, m$; j is the output unit of the neural network, $j = 1, \ldots \ldots, n$; k is the hidden unit, $k = 1, \ldots, p$; ω_{ki} is the weight coefficient between neuron i in the input layer and neuron k in the hidden layer, and ω_{jk} is the weight coefficient between neuron j in the output layer and neuron k in the hidden layer.

2. Correlation index R_{ij}:

$$R_{ij} = \left|\frac{1-e^{-y}}{1+e^{-y}}\right| \qquad y = r_{ij} \tag{22}$$

3. Absolute impact factor S_{ij}:

$$S_{ij} = \frac{R_{ij}}{\sum_{i=1}^{m} R_{ij}} \tag{23}$$

The absolute influence coefficient S_{ij} can be used to evaluate the influence of different input units on the output result, with a higher value of S_{ij} for an input unit indicating a greater influence on the result.

Input layer and hidden layer connection weights (ω_{ki})									
ω_{11}	ω_{12}	ω_{13}	ω_{14}	ω_{15}	ω_{16}	ω_{17}	ω_{18}	ω_{19}	ω_{110}
0.247	−0.263	−0.240	−0.127	−0.282	−0.263	0.266	−0.232	0.284	0.288
ω_{21}	ω_{22}	ω_{23}	ω_{24}	ω_{25}	ω_{26}	ω_{27}	ω_{28}	ω_{29}	ω_{210}
−0.662	0.305	0.664	0.173	0.241	0.303	−0.342	0.665	−0.245	−0.241
Hidden layer and output layer connection weights (ω_{jk})									
ω_{11}	ω_{12}	ω_{13}	ω_{14}	ω_{15}	ω_{16}	ω_{17}	ω_{18}	ω_{19}	ω_{110}
0.792	−0.177	−0.807	−0.043	−0.303	−0.183	0.131	−0.814	0.273	0.276
ω_{21}	ω_{22}	ω_{23}	ω_{24}	ω_{25}	ω_{26}	ω_{27}	ω_{28}	ω_{29}	ω_{210}
0.355	−0.243	−0.331	−0.059	−0.348	−0.246	0.198	−0.310	0.332	0.335
ω_{31}	ω_{32}	ω_{33}	ω_{34}	ω_{35}	ω_{36}	ω_{37}	ω_{38}	ω_{39}	ω_{310}
−0.495	−0.296	0.500	−0.100	−0.328	−0.297	0.283	0.503	0.345	0.353
ω_{41}	ω_{42}	ω_{43}	ω_{44}	ω_{45}	ω_{46}	ω_{47}	ω_{48}	ω_{49}	ω_{410}
−0.463	−0.259	0.461	−0.150	−0.030	−0.255	0.319	0.456	0.081	0.074
ω_{51}	ω_{52}	ω_{53}	ω_{54}	ω_{55}	ω_{56}	ω_{57}	ω_{58}	ω_{59}	ω_{510}
−0.267	0.236	0.259	−0.328	0.527	0.245	−0.104	0.250	−0.560	−0.520
ω_{61}	ω_{62}	ω_{63}	ω_{64}	ω_{65}	ω_{66}	ω_{67}	ω_{68}	ω_{69}	ω_{610}
−0.167	0.346	0.158	0.492	0.419	0.143	−0.597	0.157	−0.445	−0.442

Table 10.
Neural network weights of the intelligent design model.

Absolute influence coefficient S_{ij}	Scheme 1	Scheme 2	Scheme 3	Scheme 4	Scheme 5	Scheme 6
S_{ij} of plant span	0.341	0.391	0.293	0.058	0.423	0.444
S_{ij} of strength-stress ratio	0.659	0.609	0.707	0.942	0.577	0.556

Table 11.
Weight of the support scheme for the ratio of span and strength stress.

5.2 Discussion of support scheme impact factors

The weights of the neural network model in Section 4 are shown in **Table 10**.

The influence weights of the plant span and strength-stress ratio on the 6 system anchor support schemes are shown in **Table 11**.

As seen from **Table 11**, the weights of the strength-stress ratio on the results in the intelligent support design model are greater than the weights of the plant span. Therefore, when considering only the plant span and the strength-stress ratio, the variation in the strength-stress ratio has a greater influence on the choice of the anchor support scheme for the underground plant.

6. Conclusion

1. Anchor bolts or anchor cables can provide additional cohesion increments to the surrounding rock, and the support strength reflected by the anchor shows a

certain functional relationship with the strength-stress ratio and plant span. Through the statistical method of least squares fitting, four empirical formulas are proposed for the strength-stress ratio K_σ of anchor bolts and anchor cables and the plant span B.

2. For both anchor cable support and anchor bolt support, there are intervals where the strength of the support increases at a rapid rate. For anchor bolt support, the required support strength increases rapidly when the strength-stress ratio $K_\sigma \le 3$ and the underground plant rock is in a very high-stress state; for anchor cable support, the required support strength increases significantly when $K_\sigma < 4$.

3. Based on the empirical fitting formula, a dimensionless support strength index I_b is proposed, which can visually characterize the relationship between the designed support strength and the support strength obtained from statistical analysis. The support index can be used as a reference for support design.

4. Based on the theory of BP neural networks, an intelligent design model for the anchor support of underground plant systems is proposed. Using MATLAB as the development language, the function of obtaining the system anchor support scheme by inputting the plant span and strength-stress ratio of the underground plant is realized.

5. The Huangjinping, Ertan, and Baishan hydropower stations were selected as engineering cases with high, medium, and low in situ stress conditions to verify the feasibility of the intelligent design model. Compared with actual projects, the intelligent design model provides a safer and more reliable support scheme for Huangjinping and Ertan hydropower stations.

6. With the help of the support strength index concept, the support strength of the support scheme suggested by the intelligent design model was compared with that of the scheme used in actual engineering. The results show that the support scheme suggested by the intelligent design model is safer and more stable and can still achieve the desired design effect under high in situ stress conditions.

7. By calculating the absolute influence factor S_{ij}, the weights of the strength-stress ratio and the plant span for the selection of different support schemes were obtained. Based on the calculation results, the strength-stress ratio has a greater influence on the choice of the system anchor support scheme when only the plant span and strength-stress ratio are considered. This method provides a new idea for studying the influence of different factors on the choice of system anchor support scheme.

Author details

Jianhai Zhang[1], Tianzhi Yao[1*], Li Qian[1], Zuguo Mo[1], Yunpeng Gao[1], Fujun Xue[1], Chenggang Liao[2] and Zhong Zhou[2]

1 State Key Laboratory of Hydraulics and Mountain River Engineering, College of Water Resources and Hydropower, Sichuan University, Chengdu, China

2 Power China Chengdu Engineering Corporation Limited, Chengdu, China

*Address all correspondence to: sixonego@163.com

References

[1] Ye W. Monitoring and analysis of surrounding rock stability of underground powerhouse caverns of hydropower stations. Low Carbon World. 2019;**9**(12):118-119

[2] Liu WJ. Action mechanism and applications of surrounding rock reinforcement by grouting anchor. Science Technology and Engineering. 2017;**17**(20):270-276

[3] Ge HW, Liang YC, Liu W, Gu XJ. Applications of artificial neural networks and genetic algorithms to rock mechanics. Chinese Journal of Rock Mechanics and Engineering. 2004;**23**(9):1542-1550

[4] Su GS. Study on Stability Analysis and Intelligent Optimization for Large Underground Caverns under High Geostress Condition. Beijing, China: Chinese Academy of Sciences; 2006

[5] Zhang QY, Xiang W, Zhu WS. Application of 3-D elastoplastic damage model with bolts in Xiluodu underground power house. Chinese Journal of Computational Mechanics. 2000;**17**(4):475-482

[6] Huang ZJ, Sheng Q. Elastoplastic analysis on rockmass of the underground house in Jiangkou hydropower station. Chinese Journal of Rock Mechanics and Engineering. 2002;**21**(6):2098-2102

[7] Cai B, Huang ZP, Liu YK. Rock mechanics experimental study of Jiangkou hydropower station. Yangtze River. 2001;**32**(3):53-55

[8] Zhang Q, Li L, Lai DP, et al. Application of geological engineering theory to study and construction of Xiaolangdi underground plant. Water Resources and Hydropower Engineering. 2001;**32**(11):28-30

[9] Chai ZY, Xiong W, Wang Y. Analysis of the support and deformation stability of surrounding rock in the excavation of Xiaolangdi underground power house. Hongshui River. 2002;**21**(2):36-38

[10] Zhou HM, Ding XL, Sheng Q. Viscoelastic numerical simulation and stability evaluation on excavation process of underground power station of TGP. Journal of Yangtze River Scientific Research Institute. 2002;**19**(1):31-34

[11] Yuan CH. Finite Element Analysis of the Stability of the Block Wall at the Downstream of the Three Gorges Underground Powerhouse. Beijing: China University of Geosciences; 2006

[12] Xiao SR. Research on key blocks in the surrounding rock mass of underground powerhouse, the three gorges project. Hydrogeology and Engineering Geology. 2005;**32**(3):15-18

[13] Mei SH, Sheng Q, Feng XT, Liu LP, Zhao HB. Back analysis of 3D in-situ stress field of underground powerhouse area of longtan hydropower station. Chinese Journal of Rock Mechanics and Engineering. 2004;**23**(23):4006-4011

[14] Sun KC, Sun ZY. Stability analysis of surrounding rock mass of underground powerhouse in Xiangjiaba hydropower station. Journal of Yangtze River Scientific Research Institute. 2006;**23**(5):29-32

[15] Deng JL, Zhao ZY, Yang YW. Underground powerhouse surrounding rock support design for Xiowan hydropower plant. Yunnan Water Power. 2007;**23**(4):55-58

[16] Zeng J, Sheng Q, Liao HJ, Leng XL. Numerical simulation with FLAC3D on construction and excavation process of underground powerhouse of Fuziling

pumped storage hydropower station. Rock and Soil Mechanics. 2006;**27**(4): 638-642

[17] Tang XH, Zhang JH, Jiang F, Zhao WG, Wu ST, Zhou GF. Study on the design of reinforcement with statistical method. Journal of Sichuan University (Engineering Science Edition). 2007;**39**:176-181

[18] Luo CW, Liu YF, Jing F. Three dimensional geostress test in underground plant area of Shuibuya project. Journal of Yangtze River Scientific Research Institute. 1999;**16**(1): 45-47

[19] Sun HY, Shang YQ, Zhang CS. Numerical modeling analysis for surrounding rockmass stability of large underground cavities. Journal of Zhejiang University (Engineering Science). 2004;**38**(1):70-74

[20] Zhang DC, Zhang YM, Chen Q, Gao YH, Song XF. Excavation and support of underground caverns at Dachaoshan hydropower station. Water Power. 2001;**12**:32-34

[21] Zhou JP, Zhang XS, Dai QX. Rock anchor beam construction technology of Dachaoshan hydroelectric plant. Design of Hydroelectric Power Station. 2000; **16**(4):18-22

[22] Xue YJ, Chen SH. Two-stage analysis of geostress fieled for underground chamber area of Pubugou project. Chinese Journal of Rock Mechanics and Engineering. 2006;**25**(9):1881-1886

[23] Jiang F. Analysis of Surrounding Rock Stability of Jinping Underground Hydropower Station. Chengdu: Sichuan University; 2006

[24] Li L, He JD, Yu T, Fan JW. Optimization design of the longitudinal axis of underground powerhouse. Journal of Sichuan University (Engineering Science). 2003;**35**(3):34-37

[25] Zhang EB, Zhang JH, Zhao WG, Zhang LM. Study on Xiluodu hydropower station right bank underground powerhouse chamber complex stability and shear zone location sensitivity. Design of Hydroelectric Power Station. 2007;**23**(1):10-14

[26] Yang K, Zhang LX, Li ZK. New method for calculating geostresses in FEM analysis of underground houses. Chinese Journal of Rock Mechanics and Engineering. 2002;**21**(11):1639-1644

[27] Bing D, Yang K. Discussions on some general problems in 3-D non-linear FEM analysis of grouped cavern. Journal of Wenzhou University. 2005;**28**(2): 42-48

[28] Li SC, Chen WZ, Zhu WS. Study on stability and anchoring effect of jointed rock mass of an underground powerhouse. Rock and Soil Mechanics. 2003;**24**(4):510-513

[29] Zhang EB, Zhang JH, Zhou Z, Tang ZY, Wang RC. Study of unstability and support of surrounding rock in some sections of the Jinping river diversion tunnel. Yunnan Water Power. 2006; **22**(4):40-44

[30] Tang XH, Zhang JH, Zhang EB, Chen LM. Study on the cavern surrounding rock stability of the underground powerhouse of the Xiluodu hydropower station. Yunnan Water Power. 2007;**23**(1):33-37

[31] Wang JH, He RP, Wei YY. Stability analysis for surrounding rock mass of Dagangshan underground powerhouse during construction excavation. Journal of Yangtze River Scientific Research Institute. 2014;**31**(11):97-101

[32] Wu K, Sheng Q, Zhang YH. Numerical simulation analysis of the excavation process of Dagangshan underground carven group. Water Power. 2009;**35**(7):20-23

[33] Cheng LJ, Tan KQ, Zhang ZJ. Study on composite deformation failure characteristics of surrounding rocks in underground powerhouse at Changheba hydropower station. Sichuan Water Power. 2016;**35**(01):29-31+140

[34] Wang YX. Design of underground powerhouse of Lubuge hydropower station. Water Power. 1998;**12**:44-48

[35] Liu HY, Zhang QY, Lin B, Yu XY. Stability analysis of underground powerhouse cavern group of Shuangjiangkou hydropower station. Yellow River. 2009;**31**(08):107-109

[36] Du FX. Study on Surrounding Rock Stability of Large-Span and High-Side Wall Underground Powerhouse of Nuozhadu Hydropower Station. Kunming, China: Kunming University of Science and Technology; 2009

[37] Sun F, Shi XQ. Design of BP neural network based on MATLAB. Computer and Digital Engineering. 2007;**35**(08): 124-126+202

[38] Sun HJ, Wang XH. Determination of the weight of evaluation indexes with artificial neural network method. Journal of Shandong University of Science and Technology (Natural Science). 2001; **20**(03):84-86

Chapter 6

The Karst Protection Foundations Design

Natalia Gotman

Abstract

In this chapter, two karst protection methods are analyzed: the structural karst protection method, when designed structures of the underground part prevent the development of the bearing structures strains while karst deformations occur; the geotechnical karst protection method, when the design of a foundation includes a protection geotechnical screen in the base, which excludes or essentially decreases the negative influence of karst development on the bearing structures of the building. The results of numerical calculations are analyzed for different types of foundations. The advantages of the model of the variable coefficient of subgrade reaction are discussed. It is proposed to determine the coefficients of subgrade reaction (stiffness of the piles) around the karst cavity by decreasing these coefficients in relation to the same coefficients, calculated using standard methods (i.e., without karst deformations). Analytical solutions for different types of foundations are presented. Analytical solutions for designing bridge support pile foundations in karst areas are proposed. The correct design model and criteria for stability evaluation of the arch over the cavity are selected and the method for estimating the thickness of an effective karst protection geotechnical screen is proposed.

Keywords: structural karst protection, geotechnical karst protection, numerical modeling, foundations, analytical solutions

1. Introduction

The methods for analysis and design of the foundation of buildings and structures in karst areas depend on the complexes of the karst protection measures used. Two possible protection options are suggested:

- the creation of such a constructive scheme of the underground part of a building or structure that will not allow the forces of the bearing structures to exceed the permissible values;
- the installation of a protective geotechnical screen, either at the base of the foundation or above the karstic soils, will eliminate or substantially reduce the negative influence of karst development on the bearing structures.

The choice of protection against karst deformations is determined by the level of karst danger. In Russian Codes (SP 22.13330.2011), the two indicated options are assigned to the corresponding groups of measures of karst protection (structural and geotechnical, respectively) and the requirements for analysis of the karst protection foundations are significantly different.

As the results of the investigations of the karst deformations regularities [1–4], and in accordance with the accepted classification the deformations can be divided into the holes and subsidences (**Figure 1**).

The foundation calculation, providing karst deformations in a base, is the most advisable with due regard for the building and base interaction by the numerical modeling. The most simple decision in the given case is the modeling of the karst hole under the foundation base according to the hole dimensions. In places of the formation of the hole, the soil "goes out" from the foundation base and the load is redistributed to the adjacent parts where the contact of the foundation with the soil is provided. When modeling the building and base interaction with the karst hole, the choice of the base model and the determination of its initial parameters are of great importance.

The more important question is the bridge support pile foundation in the karst areas design. The design of the support pile foundation can be carried out in accordance with the comparison of the load-bearing capacity of the piles and the load transferred to the pile. Studies of the pile behavior during the formation of a karst hole in the base [5–8] show that additional vertical and horizontal loads are transferred to the piles. The additional vertical loads are considered as "negative friction" and it is indicated that they must be taken into account. However, so far there have been no proposals to define such additional loads either in the normative or scientific literature.

The numerical studies for defining additional loads on the bridge support foundation piles during the karst cavity formation in the soil under the pile bottoms, depending on the distance to the karst soil and the karst cavity predicted size, are effective. As a result of the studies performed, the regularities of changes in the additional load transferred to the pile, depending on the variable parameters, are established and formulas for calculating the support foundation piles above the karst cavity are proposed.

2. Requirements for analysis of the foundation as a constructive measure of karst protection

The purpose of constructive measures of karst protection is to prevent the destruction of the structure when karst deformations occur at the base of the foundation. These measures are designed on the basis of the analyses that ensure a

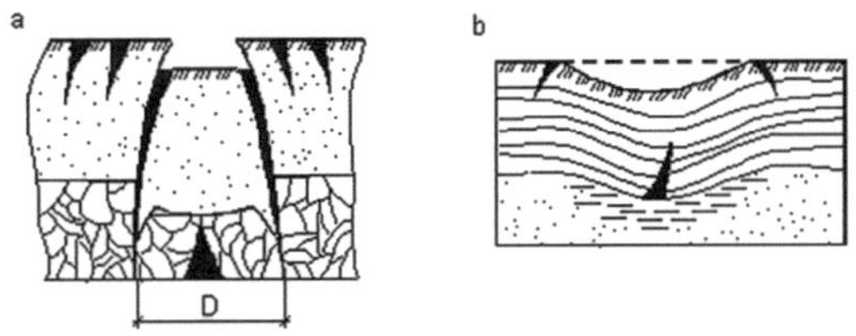

Figure 1.
Karst deformation types: a – Karst hole; b – Surface subsidence.

sufficient load-bearing capacity of the foundation and above-foundation structures to accept the additional loads that arise when karst deformations occur in the base. This is usually achieved in two ways:

- by carrying out analysis of the foundation in conjunction with the above-foundation structures for the case of the karst deformation occurrence with the specified parameters;
- by cutting through the karst soils and supporting the foundation on monolithic rocks.

Examples from the practice of design and construction show that the foundations designed for the karst deformations occurrence protects the building or structure from destruction when the karst processes in the base are activated.

However, the inclusion of a foundation that provides effective karst protection of a building can only be guaranteed if it is designed on the basis of calculated positions and initial data corresponding to the nature of the development of karst deformations. The main initial data, in this case, are the design parameters of karst deformations. The design parameters of karst deformations are determined (predicted) depending on their type.

There are three types of karst and suffusion deformations development:

- "Hole", when the karst cavity develops in karst soils and "floats" under the foundation base;
- "Subsidence" as the result of the karst and suffusion processes development in the cover mass;
- "Local subsidence", when the karst cavity develops in karst soils or the cover layer, but does not "float" under the foundation base.

The decision which kind of karst deformations is critical is determined by the soil conditions and design features of the projected building or structure. The most dangerous variant of the development of deformations is accepted for design.

For shallow buildings or structures, it is advisable to perform calculations for the occurrence of a karst hole under the foundation base (the design diameter of the karst hole is taken as the design parameter of the karst deformation) or for the formation of a cauldron with the specified parameters.

For buildings or structures with the underground part, the most dangerous can be a karst deformation of the type "local subsidence", since the foundation is approaching karst soils and the growth of the cavity in them, even if the stability of the arch is maintained, can cause significant additional forces in the bearing structures of the underground part. At the same time, the size of the karst cavity can be adopted as the design parameter of karst deformation, for which its arch is stable. **Figure 2** shows an example of determining the size of such a cavity. In this case, the mathematical modeling of the karst cavity growth is performed using the finite element calculation with elastic–plastic model of the soil by eliminating the weakened zones (zones of the local loss of stability) around the karst cavity while maintaining the constant control of the equilibrium conditions of the arch. The growth of the cavity occurs before the maximum value of its diameter is attained, at which the

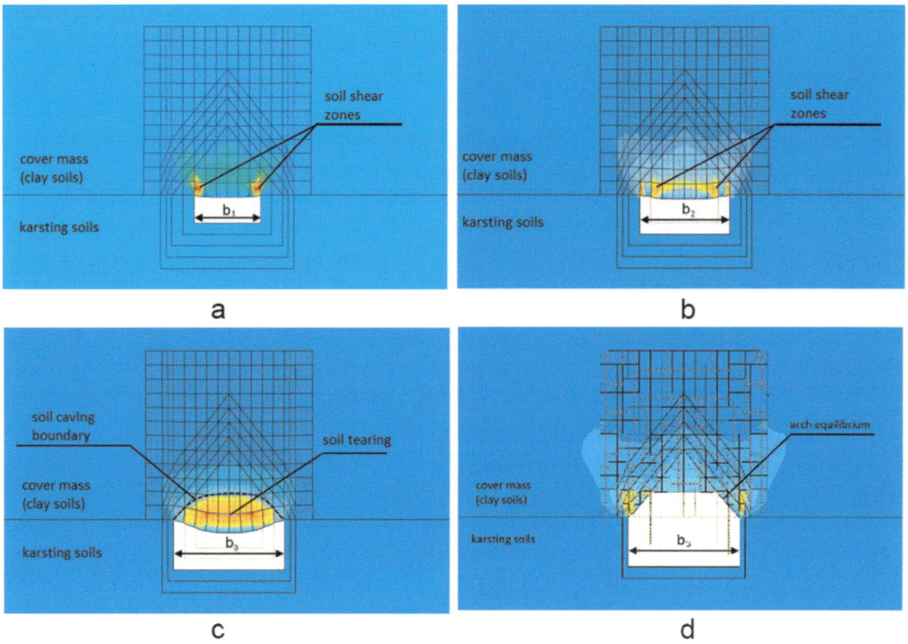

Figure 2.
The zones of equal shear strains: a, b, c – The width of the cavity is b_1, b_2, b_3, respectively (b_1 b_2 b_3); d- the cavity width (b_3) due to occurrence of the equilibrium condition of the arch.

equilibrium condition of the system is satisfied in the pre-limit state of the cover mass soil. **Figure 2** shows the lines of equal soil shear strains with the cavity width increase from value b_1 to b_3 in karst soils.

To design reliable and economical foundations, it is important to take the effect of the occurrence of karst deformations on the stress–strain state of the base and bearing structures into account. Taking into consideration the fact that the geometric dimensions of karst cavities in karstic rocks are not strictly defined, and the modeling of karst occurrence at the foundation base of a building or structure cannot guarantee the reliability of the results of the foundation analysis, the simplest solution is to model a karst hole under the foundation base in accordance with the dimensions determined by the statistical - probabilistic methods. At the same time, in the places of the formation of the hole, the ground "leaves" from under the foundation base, and the load is redistributed to adjacent areas, in which there is a contact of the foundation with the base. The modeling of the base behavior when karst deformations occur under the foundation base is possible using both elastic–plastic models of the base and the contact model.

The contact base model or the model of the variable coefficient of subgrade reaction, compared with the other base models is the simplest and most understandable for the practicing engineer. It allows both the heterogeneity of the base and its real distribution ability to be taken into account. The use of this model for the foundation in numerical modeling of the building (structure) and the base interaction also allows to reduce the order of the system of equations compared to the elastic and elastoplastic models of the base and, accordingly, the analysis errors.

Practical design experience of Russian engineers-researchers confirms the efficiency of a *combined approach* used in the foundation's analysis while karst deformations occurrence is simulated using an elastoplastic soil model for calculating stresses, deformations and coefficient of subgrade reaction of the base. The most effective way to determine the coefficients of subgrade reaction is to use the lowering coefficients with respect to the coefficients of subgrade reaction defined by standard methods without taking karst deformation into account [5].

With karst deformations of a "hole" or "subsidence" type, the compliance of the base is reduced due to the de-compaction of the soil around them with the load increase on these areas in the first case, and with the weakening of the base and unloading of the neighboring stronger sections in the second one. Therefore, it is suggested to determine the coefficients of subgrade reaction (pile stiffness coefficients) for the areas around the karst hole K_h by taking into account the decreasing coefficients ξ with respect to the coefficients of subgrade reaction (pile stiffness coefficients) K defined by standard methods without taking karst deformation into account:

$$K_h = {}^{K}\!/_{\xi}. \tag{1}$$

Based on the results of numerical and field studies, methods for determining of the coefficients ξ for raft, pile-raft, and pile strip foundations have been developed.

Analysis of the *raft foundation* on the karsted area is usually performed for the karst deformation of a "hole" type when the diameter of the karst hole is taken as the design parameter. In this case, the coefficient of subgrade reaction within the boundaries of the karst hole is equated to zero, and outside these boundaries, it decreases with approaching the hole.

For a building or structure with a developed underground part, such an approach may be erroneous and lead to unpredictable deformations of the base and stresses in the foundation sections, since the karst cavity in the karst soils may be of a larger diameter than the karst hole "floating" as a result of the cavity arch failure. At the same time, due to the proximity of the foundation base to the karst soils, the local subsidence of the base above the cavity will provoke greater forces in the foundation sections than the karst hole under the foundation base of a smaller diameter. Therefore, in this case, it is suggested to take the diameter of the karst cavity in the karst soils (d_p) as the design parameter of karst deformations. It is the maximum diameter of the karstic cavity when the soil cover mass is stable and the cavity does not "float" to the surface in kind of a hole, but there occurs local subsidence of the base above the cavity [9]. As a result of 3D finite element calculation with the elastic–plastic model of the soil, a method for analysis of the coefficient of subgrade reaction for the raft foundation base of a buried building, has been developed. This method allows to determine the decreasing coefficient ξ with respect to the coefficient of subgrade reaction determined without regarding cavity occurrence by any known methods:

$$\xi = \frac{h_k - h_f - \alpha(d_p - 3)}{h_k - h_f + \beta(d_p - 3)}, \tag{2}$$

where h_k is the depth of the karsting soils; h_f - deepening of the foundation; d_p is the cavity diameter; coefficient α = 0,871-0,0261 t; coefficient β = 1.2691–0.4163 t; t is the thickness of the foundation slab; all units are given in meters.

As shown in **Figure 3** the subgrade reaction coefficient and pressures under the raft base for the occurrence of karst cavity of the design diameter d_p decrease. The radius of the zone for reducing the coefficient of subgrade reaction (R) is determined by the formula:

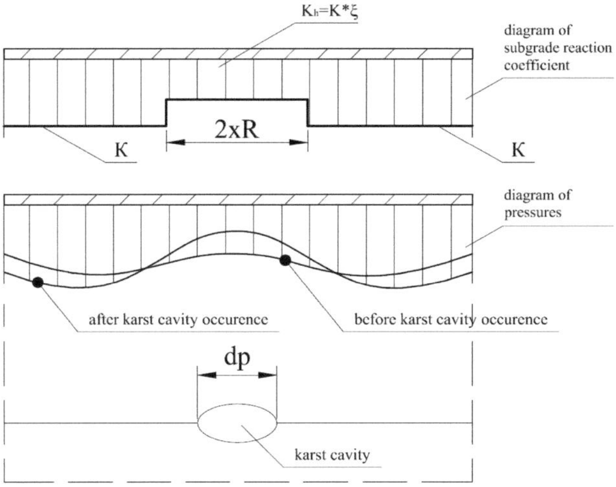

Figure 3.
A schematic of the subgrade reaction coefficient and pressures under the raft base for the occurrence of karst cavity of the design diameter d_p.

$$R = \sqrt[4]{\frac{16\,Et^3\,\beta\,(d_p - 3)S}{3P(5+\mu)(1-\mu)(h_k - h_f)}}, \tag{3}$$

where E and μ are the deformation modulus and the Poisson's ratio of the raft concrete, respectively, P is the pressure in the raft base; S – the settlement under the raft base center, defined before the karst cavity occurrence.

Analysis of *the pile-raft foundation* in karsted areas is usually performed for karst deformations of a "hole" type. The stiffness ratio of the piles is equated to zero within the boundaries of the karst hole, and outside these boundaries, it is assumed to be constant and is determined by the standard methods, that is, without taking into account the formation of a karst hole.

Due to the peculiarities of the pile-raft foundation behavior, namely, the effect of pile pre-stressing in the soil, a situation is possible when the soil mass, stabilized with piles, accepts stresses of karst deformations and the karst cavity under the pile tips does not develop to the foundation base. In this case, karst deformations should be considered as "local subsidence". In this case, the forces in the raft sections and, accordingly, the reinforcement of the raft, can be significantly reduced. Considering these features of the pile-raft foundation behavior, a method was developed for the analysis of the stiffness coefficient of the pile foundation above the karst cavity located under the pile bottoms. Analytical solutions were obtained to determine the pressures in the base and the settlements of the raft above the karst cavity [5, 8]. By the results of the analytical investigations using the linear-elastic approach, the method of calculation of the pile deformability ratio above the karst cavity is developed. The stressed-deformed state of the base with the full design column load is analyzed at the moment of the karst cavity formation under the pile bottoms. The pile compression in soil and the extra radial stresses along the pile shaft due to adjacent pile loads are taken into account.

As it is shown in **Figure 4** the pile design scheme above the karst cavity is characterized by the radial stresses σ_r and the friction force f along the lateral surface. The axisymmetrical problem of the radial stresses σ_r distribution is solved when each pile in a pile field is loaded with a load P.

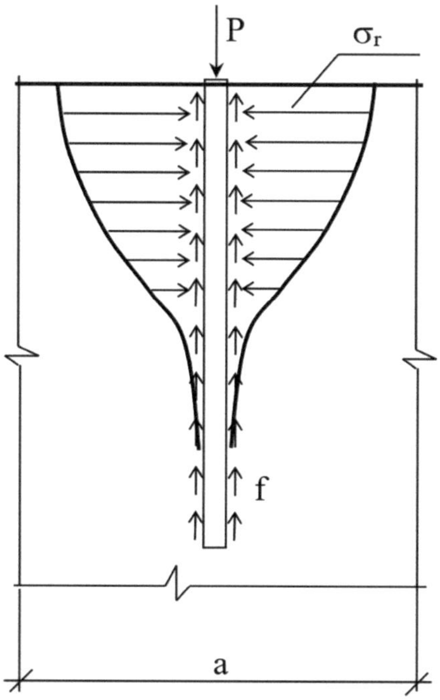

Figure 4.
Pile design scheme above karst cavity.

$$\sigma_r = \frac{1}{r}\frac{d\varphi}{dr}, \tag{4}$$

where φ is defined by the equation of the deformations compatibility, the general integral of which is the function:

$$\varphi = C_1 + C_2 Inr + C_3 r^2 + C_4 r^2 Inr. \tag{5}$$

The coefficients C_1, C_2, C_3 (Eq.5) are defined by the equilibrium of forces around the piles with the total number of piles m and the distance to the neighbor piles b_i:

$$\int_L T_\sigma d\sigma - \sum_{i=1}^{m} b_i \int_L \varphi_i(\sigma) d\sigma = 0. \tag{6}$$

The values of T_σ are defined according to R. Mindlin solution

$$T_\sigma(r = a/2) = \frac{P(L-z)}{L\pi(1-\upsilon)}\left[\frac{4z(1-2\upsilon)}{R^3} + \frac{2(1-\upsilon)(1-2\upsilon)}{R(R+2z)} + \frac{6z^3(1-4\upsilon)}{R^5} - \frac{3r^6}{R^7}\right],$$

$$R = \sqrt{r^2 - 4z^2}.$$

As the result of the solution (Eq. 6), the coefficients C (Eq. 5) are defined as the functions of pile length, pile spacing, pile cross-section, and distance Z_i from the soil surface. With the coefficients, C the solution for the evaluation of the radial stresses σ_r from unit loads on the pile is obtained (Eq. 4, Eq. 5).

The condition, when the piles do not "move" in soil and karst deformations should be considered as "local subsidence", is evaluated by the expression:

$$P_{lim} < \left(\sum_{i=1}^{n} \tau_{i,lim} U h_i \right) - \gamma_{av} L a^2, \tag{7}$$

where P_{lim} is the pile limit load above the karst cavity and is evaluated as $P_{lim} = pa^2$; p is the pressure transmitted to a raft base; a is the pile spacing; U is the pile perimeter; h_i is the length of the ith section; n is the number of sections by the pile length; L is the pile length; γ_{av} is the weighted average value of soil density; $\tau_{i,lim}$ is the soil specific resistance by Coulomb accounting the stress σ_r and the friction force f

$$\tau_{i,lim} = c_i + tg\varphi_i (P\sigma_{ri} + \gamma_i z_i \beta), \tag{8}$$

where P is the given pile load (by the calculation of the pile field in conditions of normal operation when the karst holes are not formed); c_i, φ_i, γ_i are the specific cohesion, angle of inner friction, and the soil density of the considered layer, respectively; σ_{ri} is the stress of pile-soil compression due to the unit loads at the distance z_i from the soil surface; β is the lateral pressure coefficient.

To define the pressure (p) transmitted to a raft base, a problem for the foundation piled raft is solved for the case of a karst cavity under the pile bottoms. The foundation raft is considered as the plate of the infinite radius on the combined base with the karst cavity of r_k radius.

The solutions of Russian scientist Korenev B.G. are used to evaluate the pressure in the raft base and the settlements of the raft base in Bessel functions [5]:

$$p_r = \frac{N}{2\pi} \int_0^\infty \frac{\gamma J_0(\gamma r) d\gamma}{1 + \frac{D}{k_0}\gamma^4 + cD\gamma^4}, \tag{9}$$

$$\omega_r = \frac{N}{2\pi} \int_0^\infty \frac{\left(\frac{\gamma}{k_0} + c\gamma\right) J_0(\gamma r) d\gamma}{1 + \frac{D}{k_0}\gamma^4 + cD\gamma^4}, \tag{10}$$

as well as Hankel conversion for the function c evaluation:

$$c = 2\pi \int_0^\infty rK(r) J_0(\gamma r) dr, \tag{11}$$

where k_0,k - pile deformability ratio (k_0), the bed coefficient of the elastic half-space (k); $J_0(\gamma r)$- the Bessel integral; D- plate cylindrical stiffness $Eh^2/12\ (1-\nu^2)$; E-concrete deformation modulus; h-plate thickness; ν- Poisson's ratio, γ- is defined from the boundary conditions due to the karst cavity of r_k radius formation.

The function $K_{(r)}$ corresponding to the settlement surface when karst cavity formation under the pile bottoms is taken as:

$$K_{(r)} = \frac{B}{2\pi r} \exp(-\delta r). \tag{12}$$

The parameters B, δ are defined from the boundary conditions due to the karst cavity of r_k radius formation.

By the results of the calculations of the improper integrals of Bessel's function, the pressures in the raft base (p_r), raft settlements above the karst cavity (ω_r) due to the unit load depending on the radius of the karst cavity under the pile bottoms (r_k), raft thickness (h), and the pile deformability ratio (k_O) are defined.

Using the analytical dependence of the pressure in the raft base on the pile deformability ratio above the hole and solving the inverse problem, the pile deformability ratio (k_O) is calculated that corresponds to the given pressure (p_r).

Analysis of the *pile strip foundation* with karst deformations of the "hole" type is usually performed by mathematical modeling of the foundation on an unevenly deformed base. For modeling of the base and foundation, in this case, it is advisable not to complicate the calculation model, but, on the contrary, to apply the simplified models. Such a calculation model of the pile strip foundation base when a karst hole occurs is the contact base model, according to which the piles around the karst hole behavior is modeled by constraints of finite stiffness, defined as the stiffness ratio of the piles. The piles under the karst hole are excluded. Based on the results of the experimental and theoretical studies of the stress–strain state of the "pile strip foundation - base" interaction, the regularities of the change in pile behavior around the karst hole were obtained. As a result of 3D finite element calculation with the elastic–plastic model of the soil the analytical solution was developed to determine the stiffness ratio of piles [6]. The analytical solution allows to determine the decreasing stiffness ratio ξ with respect to the stiffness ratio, determined without taking the karst hole into attention. The decreasing stiffness ratio ξ is determined depending on the pile length (L), the hole depth (H), the distance from the pile to the hole boundary (B):

$$\xi = 1 + \frac{0.041H^4}{L^2(B^2 + 0,04H^2)}. \tag{13}$$

3. Requirements to analysis as a geotechnical measure of karst protection

One of the most effective geotechnical karst protection measures is the cementation of the cover soils above the karst soils. Schemes of karst protection cementation of the foundation are developed on the basis of the Russian Code (SP 22.13330.2011), which recommends cementation of the cavities and the entire thickness of the karst soils. However, in practice, these strata often reach considerable sizes (from 15 to 20 m), and their cementation to the full depth to monolithic rock soils where karst cavities do not form is not possible due to a significant rise in the cost of construction and technological problems of cementation and control its quality at great depths.

Investigations of the stress–strain state of the artificially strengthened foundation base over the karst cavity [10] made it possible to establish that the most efficient method was the cementation of the soil mass in the roof of the karst soils. In this case, the additional deformations in the foundation base are minimal when cavities occurrence in karst soils. Also, soil collapse into the karst cavity is not allowed if the height of the probable collapse area above the cavity does not exceed the thickness of the artificially strengthened soil layer.

The design forecast of the possibility of soil collapse into the karst cavity is based, as a rule, on the classical view of the distribution of stresses and the mechanism of the arches formation above the karst cavities. The arch above the cavity in the equilibrium state (up to the moment of its collapse) can be considered as an area of increased stresses and deformations, the size of which is determined by the strength and deformation characteristics of the soils. When cementing a soil layer of a given thickness over the karst soils in which the cavity growth is predicted, this area depends on the thickness and characteristics of the cemented soils, as well as on the maximum predicted cavity size for the standard operating life of the building and on the building loads.

The state of the collapse process characterizes the excess of the boundary values of the tensile and compressive stresses around the cavity that can be obtained from Mohr's circle of stress. Therefore, the boundaries of the region of increased stresses and the formation of shear strains can be determined using the strength condition according to Mohr-Coulomb failure criteria, taking it as the boundary condition of rock flow and its collapse. Thus, the boundaries of a possible collapse area are defined as the locus of points at which the Mohr-Coulomb failure criteria are met.

Determination of additional load on the bridge foundation pile under karst deformation.

During the formation of a karst hole in the base of the bridge support pile foundation the additional loads are transferred to the piles. The additional vertical loads can be taken into account as the tangential stresses on the lateral surface of the piles, directed to the pile base.

3.1 Finite element research technique and justification of the accepted computational model

To develop an engineering calculation method, complex numerical studies were performed by means of mathematical modeling of the bridge support pile foundation in various geological conditions. The calculation method was based on the analysis of the design documents for bridge crossings at Moscow-Kazan HSL section. As a result of the analysis, a variable finite element calculation model with the following parameters was compiled:

- the soil mass under the rocky karst soil is represented by firm clay with the characteristics specified in **Table 1**;
- a square-shaped grillage combining 36 piles with a diameter of 1.2 m and a length of 33 m.;
- the size of the calculated area L (along X and Y axes) was determined by the condition that it did not affect the results of the calculation, the nodes at the boundaries of the area were fixed;

Characteristic name	Clay
Density, kN/m^3	18.0
Deformation modulus, MPa	25.0
Angle of internal friction, degree	20
Cohesion, kPa	80.0

Table 1.
Physical and mechanical characteristics of soils.

- the size of the calculated area H (along Z-axis) was limited by the roof of the rocky karst soils, the nodes at the boundaries of the area were fixed, except for the predicted karst cavity;
- the predicted karst cavity was represented by the absence of anchoring of the nodes (along Z-axis) within it;
- the pile-soil contact was taken into account with the help of special interface elements.
- The Finite element model section and scheme of the calculated foundation is shown in **Figure 5**.

As shown in **Figure 6** the calculations were performed by varying the following parameters:

- the distance to the rock roof (*b*): 6 m, 10 m, 14 m, 18 m, 22 m;

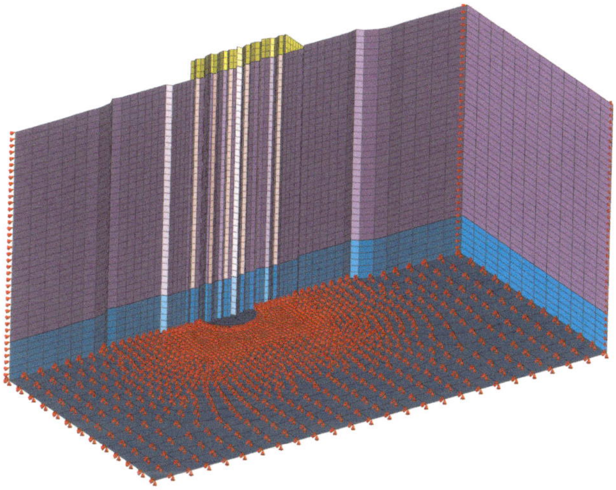

Figure 5.
Finite element model (section).

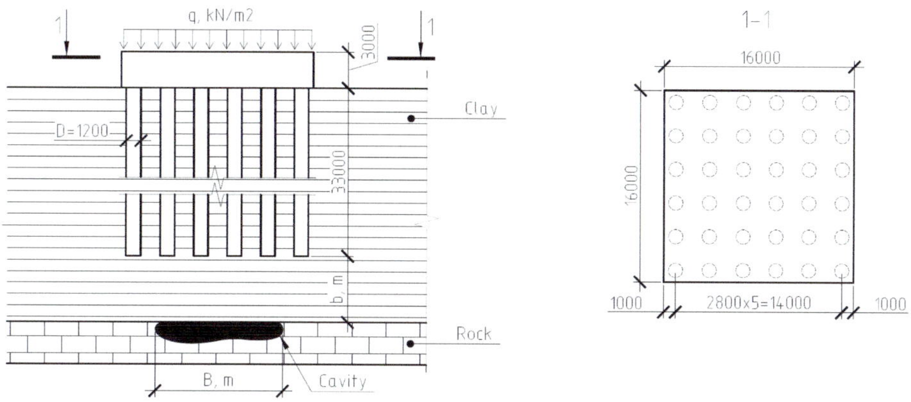

Figure 6.
The scheme of calculated foundation.

- estimated cavity size during operation (*B*): 3 m, 5.5 m, 7.8 m, 10 m;
- distributed load over the grillage top: 400 kN/m2, 550 kN/m^2 (corresponds to the load on the pile of 2800 kN and 3900 kN).

Finite element calculations were made in a three-dimensional representation with Midas GTS NX Software. Soil, grillage, and piles were modeled by three-dimensional elements. A linear-elastic model was used to model concrete. The elastic–plastic Mohr-Coulomb model was used for soil modeling with 3-dimensional finite elements. Using the strength criterion implemented in the model, it was possible to estimate the "collapse arch" size in the cover layer of the soil above the karst cavity. In this way, the "subsidence" deformation type and "failure" deformation type can be realized. The possibility of using that strength criterion was confirmed by the convergence of the calculation results with the model experiment data of the "collapse arch" formation above the cavity [10].

The calculation was performed in the following sequence:

- The initial stress–strain condition of the soil mass was determined;
- The pile foundations of the supports were calculated for the design loads under normal operating conditions and the tangential stresses on the lateral surface of the piles were determined;
- The cavity size growth in karst soils located at a given distance from the bottom of the piles was determined step-by-step and the tangential stresses on the lateral surface of the piles were defined.

During calculations, the growth of the "collapse arch" above the karst cavity was monitored. **Figure 7** shows the predicted "collapse arch" under pile bottom with

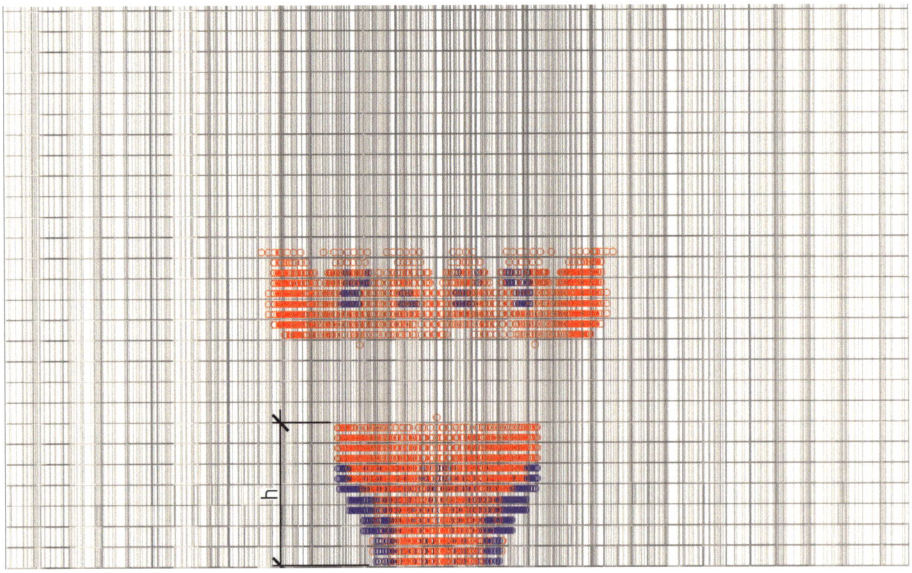

Figure 7.
Mohr-coulomb points above cavity.

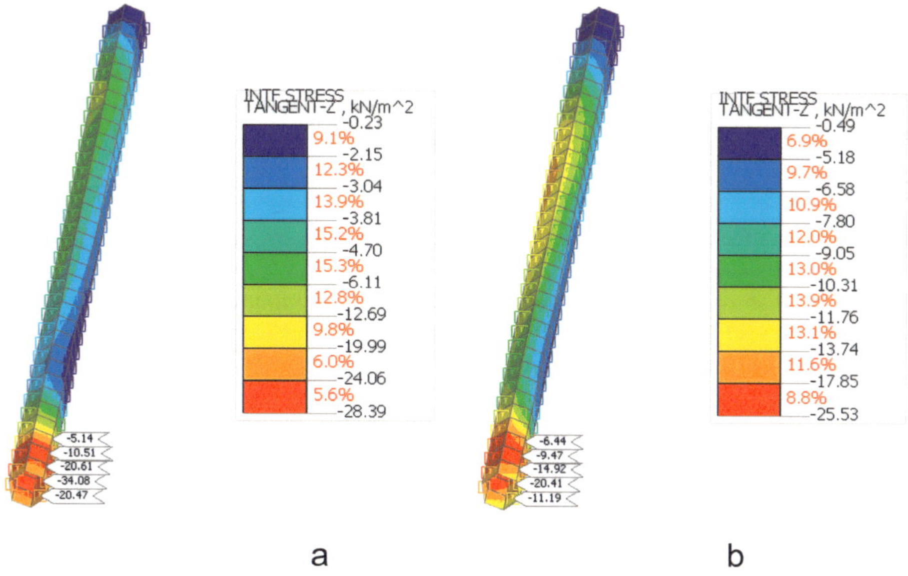

Figure 8.
Tangential stresses on the lateral surface of the pile (τ_z, kN/m2): a – Before cavity formation, b – After cavity formation.

Mohr-Coulomb points above karst cavity. Assuming the possibility of the arch development not higher the bottom of the piles, the additional load on the pile, realized at the time of the cavity formation, was determined.

The additional load was transferred to the pile at the time of the cavity formation due to the occurrence of "negative friction" on the lateral surface of the piles in their lower part. When modeling the formation of a cavity in karst soils, the occurrence of "negative friction" was determined by changing the tangential stresses on the lateral surface of the piles in comparison with the calculated ones in normal operating conditions.

Under normal operating conditions, tangential stresses on the lateral surfaces of piles increased with depth, while on the extreme and corner piles the growth began from the top of the pile (the pile was included in work entirely). In the central piles, tangential stresses developed in the lower part of the pile (due to the "compression" effect, the side surface friction of the central piles was not fully realized). Similar results of experimental and theoretical studies of piles behavior in the group were obtained in Russian and abroad [11–14].

When a cavity was formed, the soil of the cover layer subsided, which led to a change in the nature of the pile lateral surface work: the tangential stresses on the lateral surface in the lower part decreased, but along the rest of the pile length they increased. **Figure 8** shows tangential stresses on the lateral surface of the pile before cavity formation (a) and after cavity formation (b). That indicated the occurrence of the "negative friction" effect in the lower part of the piles and the inclusion of the most part of its lateral surface at the time of the cavity formation. The additional load on pile P_1, kN, was determined by the formula:

$$P_1 = u \cdot \sum \Delta\tau_{z,i} \cdot h_i, \tag{14}$$

where: u is the perimeter of the pile, m; $\Delta\tau_{z,i}$ is the change in the shear stress value on the pile lateral surface in the *considered* ith layer in comparison with the design phase under normal operating conditions, kN/m^2; h_i is the thickness of the ith soil layer in contact with the lateral surface of the pile, m. Thus, when calculating (Eq. 14) only those layers were taken into account where τ_z decreased or its direction changed.

The proportion of the increase in the load on the pile $\Delta P=P/P_1$, where P was the load on the pile under normal operating conditions, was determined. With these data, the graph for the dependence of the value ΔP on the ratio b/B was plotted. So, the additional load on the pile can be determined, having the values of the load on the pile in normal operation P, the distance from the roof of the karst soils to the bottom of the piles b, and the calculated diameter of the karst cavity B.

4. Analysis of finite element study results

After performing the variable calculations, the following results were obtained and processed:

- the position of the Coulomb-Mohr points above the karst cavity to assess the size of the "collapse arch" and control the development of that arch to the bottom of the piles (**Figure** 7);

- change of the tangential stresses on the lateral surface of the pile ($\Delta\tau_z$ kN/m^2) during the growth of the karst cavity (**Figure 8**).

As the calculation result of Coulomb-Mohr points location shown in **Figure** 7, the curves of the ratio h/b ("arch collapse" height/distance from the karts soil roof to the bottom of the pile, respectively) dependence on the ratio b/B (distance from the karts soil roof to the bottom of the pile/cavity diameter, respectively) were plotted. The defined parameters are shown in **Figures 6** and 7. The graphs for two loads over the grillage top are shown in **Figure 9**. As can be seen from **Figure 9** the possibility of

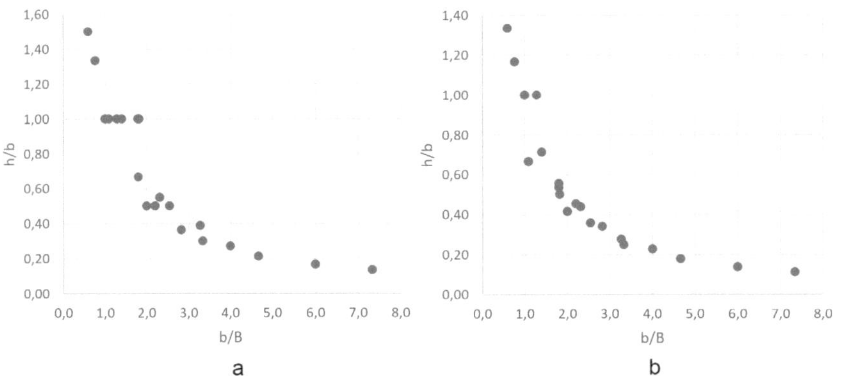

Figure 9.
Dependence of the relative distance from the predicted top of the collapsed vault to the bottom of the piles (h/b) *on the ratio of the distance from the top of karst soils to the bottom of the piles to the maximum cavity diameter* (b/B)*: A, b – Distributed load over the grillage top: 400 kN/m², 550 kN/m².*

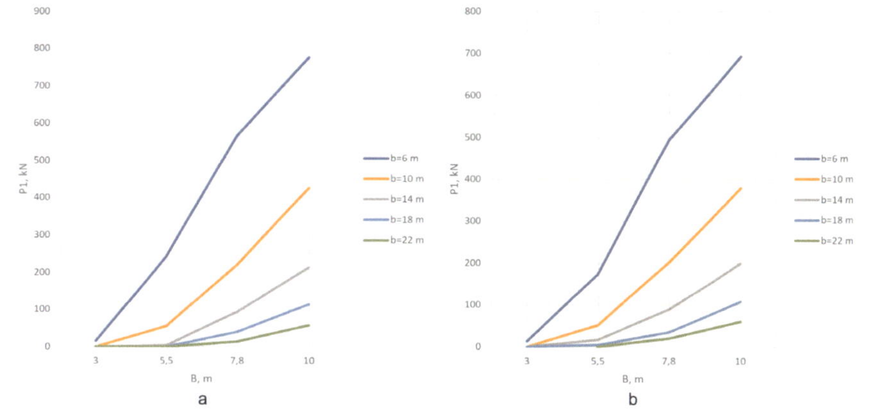

Figure 10.
Dependence of the additional load on the pile (P_1, *kN*) *on the diameter of the cavity* (B, *m*) *and the distance to the karst soil roof* (b, *m*)*: A, b – Distributed load over the grillage top: 400 kN/m², 550 kN/m².*

developing the arch above the cavity not higher than the bottom of the piles was determined by the condition $h/b \leq 1$. This condition could be met at $b/B > 1$, which limited the scope of the results of this study and the proposed solution for determining the additional load on the pile.

Figure 10 shows graphs of the dependence of the additional load on the pile (P_1, kN) on the diameter of the cavity (B, m) and the distance to the karst soil roof (b, m). These graphs are based on variable Finite element 3D calculations of the bridge support pile foundation under the karst cavity in karst soils and show that the P_1 increases with the extension of the cavity (B) and decreases of the distance to the karst soil roof (b). However, such calculations are complicated and laborious and require specialized Software to make them.

To determine the additional load on the pile (P_1) to the load on the pile under normal operating conditions (P) before the formation of the cavity, statistical processing of the calculation results was performed. It can be indicated from **Figure 11** that the relationship between P/P_1 and b/B has the form of the power function. As a result of statistical analysis the analytical dependence P/P_1 from b/B was obtained:

$$P/P_1 = 2,1e^{1,7 \cdot \frac{b}{B}}. \tag{15}$$

The value of a reliable approximation was $R^2 = 0.9194$. Thus, the additional load on the pile P_1 was determined in accordance with the design load transferred to the pile under normal operating conditions (P) before the formation of the cavity, the size of the cavity (B), and the distance from the bottom of the piles to the karst soil roof (b):

$$P_1 = P/2,1e^{1,7 \cdot \frac{b}{B}}. \tag{16}$$

5. Conclusions

1. To design the reliable and economical foundations, the practical design experience of Russian engineers-researchers confirms the efficiency of a *combined approach* used in the foundation's analysis while karst deformations occurrence based on the use of the finite-element method with elastoplastic soil

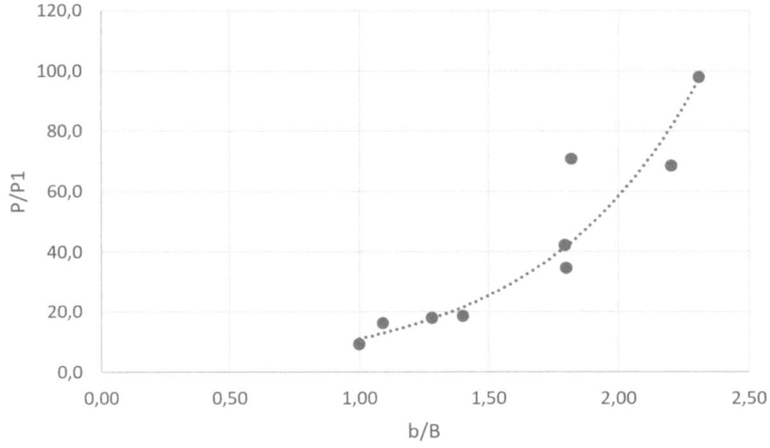

Figure 11.
Dependence of P/P1 *on* b/B *ratio.*

model for calculating stresses, deformations, and coefficient of subgrade reaction of the base.

2. The most effective way to determine the coefficients of subgrade reaction is to use the lowering coefficients with respect to the coefficients of subgrade reaction defined by standard methods without taking karst deformation into account.

3. Based on the results of numerical and field studies with the elastoplastic soil model, methods for determining the lowering coefficients for pile-raft, pile strip, and raft foundations have been developed [5, 6, 9].

4. One of the most effective geotechnical karst protection measures is the grouting of the cover mass above the karst soils. In this case, the soil collapse into the karst cavity will not occur if the height of the probable collapse area above the cavity does not exceed the thickness of the artificially strengthened soil layer. The boundaries of a possible collapse area can be defined as the locus of the Mohr-Coulomb failure criteria points [10].

5. On the basis of numerical calculations for bridge support pile foundation, it was shown that there were additional loads on the piles under karst deformations at the base of the foundation. The value of those loads for clay cover layers over karst soils was determined.

6. For bridge support pile foundation it was established that additional loads largely depended on the distance from the bottom of the piles to the karst soils roof and the cavity diameter in the karst soils. The method for calculating the additional loads on the bridge support piles depending on the size of the cavity and the distance from the bottom of the piles to the karst soil roof has been developed.

Acknowledgements

The author expresses gratitude to his graduate students Kayumov M., Davletyarov D., Evdokimov A., together with whom the research was carried out and the solutions presented in this paper were obtained.

Author details

Natalia Gotman
Russian University of Transport, Moscow, Russia

*Address all correspondence to: gotmans@mail.ru

References

[1] Tolmachyov VV. Methods of karst hazard assessment for the construction purposes: State-of-art and prospects. Geoecology. Engineering geology. Hydrogeology. Geocryology. 2012;**4**: 354-363

[2] Aderhold G. Classification of sinkholes and subsidence troughs in karst-hazardous regions of Hesse. Recommendations for assessing geotechnical risks during construction activities. Wiesbaden; 2005. p. 113

[3] Khomenko VP. Karst-suffusion processes and their prediction (Moscow Nauka): 1986. p. 93

[4] Anikeev AV. Collapses and sinkholes of subsidence in karst areas: mechanisms of formation, forecast and risk assessment (Moscow RUDN): 2017. p. 328

[5] Gotman NZ. Analysis of karst protective foundations of buildings and structures. Soil Mechanics and Foundation Engineering. 2008;**1**:20-25. DOI: 10.1007/s11204-008-0005-x

[6] Gotman NZ, Davletyarov DA. Calculation of the stiffness coefficient of piles of a pile-strip foundation for the formation of a karst hole. Soil Mechanics and Foundation Engineering. 2017;**2**:2-6. DOI: 10.1007/s11204-017-9436-6

[7] Waltham T. Sinkholes and subsidence: karst and cavernous rocks in engineering and construction. Chichester: Praxis Publishing Ltd; 2005. p. 383

[8] Iljichev VA, Gotman NZ. Evaluation of the base design parameters in the numerical modeling of a structure and a pile base interaction with karst deformations. Ljubljana, Slovenia: Proceedings of the XIII Danube-European Conference on geotechnical engineering; 2006. pp. 605-610

[9] Gotman NZ, Kayumov MZ. Calculation of the foundations of buildings with developed underground parts in karsted areas. Soil Mechanics and Foundation Engineering. 2013;**4**: 13-18. DOI: 10.1007/s11204-013-9227-7

[10] Gotman NZ. Evaluation of karst manifestation parameters by results of numerical calculations of foundations. Soil Mechanics and Foundation Engineering. 2019;**5**:2-7. DOI: 10.1007/s11204-019-09605-7

[11] Gotman NZ, Alekhin VS. Calculation of the ultimate resistance of the pile bases in pile groups. Soil Mechanics and Foundation Engineering. 2020;**1**:8-13. DOI: 10.1007/s11204-020-09632-9

[12] Randolph M. Analysis of deformation of vertically loaded piles. Journal of Geotechnical and Geoenvironmental Engineering. 1978: 1465-1488

[13] Randolph M. Design Methods For Pile Groups and Piled Rafts. Soil mechanics and foundation engineering. 1994;**13**:61-82

[14] Katzenbach R. Recommendations for the design and construction of piled rafts. Istanbul: 15th International Conference on Soil Mechanics and Foundation Engineering; 2001. pp. 927-930